SOLARIA BINARIA

SOLARIA BINARIA

ORIGINS AND HISTORY OF THE SOLAR SYSTEM

by
ALFRED DE GRAZIA
and
EARL R. MILTON

Metron Publications
Princeton, New Jersey

ISBN: 978-1-60377-096-5
Library of Congress Catalog Number: 2018902717
Copyright©Alfred de Grazia, Earl R. Milton 1984
All rights reserved
METRON PUBLICATIONS, P.O. Box 1205
Princeton, NJ, 08540-1205
http://metron-publications.org/
metronax@gmail.com

TO THE MEMORY OF

RALPH JUERGENS

τα δε παντα οιχιζει χεραυνος*

* "Lightning steers the universe"

Heraclitus, *ca.* 2500 BP,

Fragment 64

Earl Milton and Alfred de Grazia

Lake Kashagawigamog, Ontario, 1984

Earl R. Milton was born February 26, 1935 in Montreal, Canada. His academic degrees include the B.S., M.S. (in Chemistry), and Ph.D (in Chemical Physics), all from University of Alberta; he received a National Research Council Post-Doctoral Fellowship to study spectroscopy and was awarded the Chant Medal of the Royal Astronomical Society of Canada (silver) in 1960. He taught at the University of Saskatchewan in Regina, at Lethbridge Junior College, and, from 1967 to 1990, at the University of Lethbridge, where his fields were Physics, Astronomy, and Interdisciplinary Studies. He was a member of the Royal Astronomical Society of Canada and the President of its Edmonton Center for 1958-59. He gave interviews on C.N.C. and appeared on many news broadcasts and talk-shows. He was a founding member of the Society for Interdisciplinary Studies in London, England, and of the Canadian Society for Interdisciplinary Studies. He published in the *Journal of Physical Chemistry*, the *Journal of Chemical Physics*, the *Journal of the Royal Astronomical Society of Great-Britain*; in *Kronos: A Journal of Interdisciplinary Synthesis*, and in *SISR, the Society for Interdisciplinary Studies Review*. He is the author of *Recollections of a Fallen Sky* (1974). He died November 6, 1999, in Calgary, Canada.

Alfred de Grazia was born December 29, 1919 in Chicago and was a major political scientist of his generation, PhD University of Chicago, he taught at University of Minnesota, Brown University, Stanford University and New York University, and many others, in the US and abroad, authoring over thirty books in the field, as well as twenty more in fields ranging from catastrophism, and ancient history to American history and poetry. He was a pioneer in many fields, including digital archiving, as early as the 1960s. His website, grazian-archive.com, garners 110 million hits/year. He died in France on July 13, 2014.

Solaria Binaria was born of a collaboration between Alfred de Grazia and Earl Milton, which started in 1978 in Washington DC and Princeton, NJ, and was mainly carried out in Naxos, Greece, in the Spring of 1980. It was published by Metron Publications in Princeton in 1984.

Table of Contents

INTRODUCTION	*18*
PART ONE: ORIGIN AND DEVELOPMENT OF THE BINARY SYSTEM	**21**
CHAPTER ONE:	
The Solar System as a Binary	22
CHAPTER TWO:	
The Solar System as Electrical	25
CHAPTER THREE:	
The Sun's Galactic Journey and Absolute Time	36
CHAPTER FOUR:	
Super-Uranus and the Primitive Planets	50
CHAPTER FIVE:	
The Sac and its Plenum	61
CHAPTER SIX:	
The Electrical Axis and its Gaseous Radiation	67
CHAPTER SEVEN:	
The Magnetic Tube and the Planetary Orbits	75
CHAPTER EIGHT:	
The Earth's Physical and Magnetic History	82
CHAPTER NINE:	
Radiant Genesis	95
PART TWO: THE DESTRUCTION OF THE SOLAR BINARY	**107**
CHAPTER TEN:	
Instability of Super-Uranus	108
CHAPTER ELEVEN:	
Astroblemes of the Earth	118
CHAPTER TWELVE:	
Quantavolution of the Biosphere: Homo Sapiens	130
CHAPTER THIRTEEN:	
Nova of Super-Uranus and Ejection of the Moon	143
CHAPTER FOURTEEN:	
The Golden Age and Nova of Super-Saturn	159

CHAPTER FIFTEEN:
   The Jupiter Order 172
CHAPTER SIXTEEN:
   Venus and Mars 181
CHAPTER SEVENTEEN:
   Time, Electricity and Quantavolurion 194

PART THREE:

TECHNICAL NOTE A:
   On Method 199
TECHNICAL NOTE B:
   On Cosmic Electrical Charges 210
TECHNICAL NOTE C:
   On Gravitating Electrified Bodies 214
TECHNICAL NOTE D:
   On Binary Star Systems 221
TECHNICAL NOTE E:
   Solaria Binaria in relation to Chaos and Creation 227

OMNINDEX ***230***

LIST OF FIGURES

FIGURE 1: Dumb-bell Motion of Solaria Binaria 22
FIGURE 2: The Sun's Connection to the Galaxy 34
FIGURE 3: Stars around the Sun's Antapex 43
FIGURE 4: Nearby Stars in the Solar Wake 44
FIGURE 5: The Solar Antapex 47
FIGURE 6: Electron Flow from Surrounding Space into a Star-cavity 51
FIGURE 7: The Birth of Solaria Binaria 53
FIGURE 8: Material Flow Coupling the Sun, Super Uranus, and the Electrified Plenum 54
FIGURE 9: Flow of Material between the Sun and Super-Uranus under the Influence of a Self-generated Magnetic Field 55
FIGURE 10: Magnetic Toroidal Field produced by Solar Wind Current Sheet 56
FIGURE 11: Magnetif Field surrounding several Flowing Ions 57
FIGURE 12: The Planet Saturn in Ancient Indian Art 73
FIGURE 14: Decreasing Magnetic Field Strength surrounding Central Current at increasing Distances 76
FIGURE 15: Motion of Drifting Charged Particle in a Magnetic Field 78
FIGURE 16: Braking Radiation emitted by a Spiraling Electron 79
FIGURE 17: The Primitive Planets in Orbit about the Electric Arc 80
FIGURE 18: The Earth in the Magnetic Tube 84
FIGURE 19: The Earth Magnet 85
FIGURE 20: Magnetic Transactions within the Earth 87
FIGURE 21: Transaction between Solaria Binaria and the Cosmos: Dense Plenum Phase 111
FIGURE 22: Solaria Binaria as the Plenum thins and the Stars separate... 113
FIGURE 23: Explosive Eruption from Super Uranus 119
FIGURE 24: PossibleAstroblemes in Arizona 122
FIGURE 25: Meteroid Trajectories 124
FIGURE 26: Radioacivity of Fosilized Remains 135
FIGURE 27: The Suviving Land from the Age of Urania 148
FIGURE 28: The Encounter of Uranus Minor with the Earth 149
FIGURE 29: The Fractured Surface of the Earth 152
FIGURE 30: Fragmentation of Super Uranus 155
FIGURE 31: Fission of the Earth-Moon pair 156
FIGURE 32: The Chinese Craftsman God and his Paredra 163
FIGURE 33: The Churning of the Sea 165
FIGURE 34: The Golspie Stone 168

FIGURE 35: Apparent Motion of the Charged Sun about the Earth 175
FIGURE 36: The Electric Field between Mars and the Moon 192
FIGURE 37: Potential Energy Curve fo the Collision of Two Atoms 219
FIGURE 38: Electric Forces between Celestial Bodies 220
FIGURE 39: Binary Orbits of Short Period 225

LIST OF TABLES

CHAPTER THREE:

TABLE 1: Stars behind the Sun (to 25 000 years ago) 44
TABLE 2: Stars behind the Sun (from 25 000 to 75 000 years ago .. 45
TABLE 3: Stars behind the Sun (over 75 000 years past) 48

CHAPTER EIGHT:

TABLE 4: Calculated Undisturbed Decay of the Earths' Magnetization 91

CHAPTER ELEVEN:

TABLE 5: Modes of Meteoroid Encounters 125

CHAPTER TWELVE:

TABLE 6: Ages of Solaria Binaria 131

LIST OF ABBREVIATIONS IN TEXT

BP	before the present
cf.	compare
E	evolutionary (model)
EM	electromagnetic
f. (ff.)	following page(s)
fn.	footnote
Gm, Gy	gigametre, gigayear (= aeon)
ibid.	in the same place
ISEE	International Sun Earth Explorer (a space craft)
K	Kelvin
km/s	kilometres per second
ly	light year
mks	metre-kilogram'second (units)
My	megayear or million years
NMP, NRP	North magnetic (rotational pole
o.	Omnindex
op. cit.	in the work cited
Q	quantavolutionary (model)
q. v.	refer to
SB	Solaria Binaria (model)
SMP, SRP	South magnetic (rotational) pole

GUIDE TO METRIC UNITS

Distances measured in metres

Multiples of the metre, by thousands and thousandths, have special names designated by a prefix, such as micrometer and gigametre. Other metric units use the same prefixes for their multiples, like microvolts, gigaergs, etc.

Prefix	Decimal Notation	Useful to measure
nano	0.000 000 001	Atoms
micro	0.000 001	Cells
milli	0.001	type size
--	1.0	People
kilo	1 000.0	driving distances
mega	1 000 000.0	planet diametres
giga	1 000 000 000.0	star diameters
tera	1 000 000 000 000.0	Planet orbits

INTRODUCTION

Since 1924, when the theory of the expanding Universe was first expounded, the phenomena of astronomy have been viewed increasingly as intensely energetic. The notion of an explosive Universe has been abetted by the identification of novas, the discovery of the immense energy trapped in the internal structure of the atom, and the detecting of radio noises from vast reaches of space signaling events so extreme as the imploding of whole galaxies. What began as a whisper in scientific circles of the late nineteenth century has become, in late years, a shout. Yet, for reasons that can only be called ideological, that is, reflecting a constrained cognitive structure in the face of contradictory perceptions, scientific workers on the whole have not heard the "shout".

At the same time as the space and nuclear sciences have had to confront a new set of facts, the near reaches of space have been surveyed and the body of the Earth searched more thoroughly. The results confirm that the wars of the Universe have been disastrously enacted upon battlefields within the Solar System. Without exception, the planetary material that has been closely inspected exhibits the effects of extreme forces unleashed upon it. Mars, Moon, Venus, Mercury - all are heavily scarred; Jupiter and

Saturn are in the throes of internal warfare. An asteroidal belt that may be called "Apollo" represents a planet that exploded. Nor can we exclude from the common experience this scarred Earth.

Consistent with the panorama of catastrophes, and additionally supplying a new dynamic form in cosmogony, there has been developed a body of knowledge and speculation surrounding the phenomena of stellar binary systems. The first binary star orbit was computed in 1822, but not until the past few years has sufficient information become available to speak about binaries systematically. Since the first discovery, a large proportion of observed stars have come to be suspected as multiple star systems.

Moreover many cosmogonists speculate that the Solar System itself was once a binary system, or at least is now a kind of fossil binary system, with Jupiter exhibiting star-like traits. It may be pointed out, for instance, that the distance between the principal bodies of the Solar System is comparable with the distances between the separate components in many binary systems.

Hence it becomes logical that a cosmogony of the Solar System should be modeled after the theory that it was, and is, a binary system, a Solaria Binaria, accepting and applying for the purpose of the model what is known and thought about the observed stellar binaries elsewhere.

The explosive or catastrophic Universe poses basic problems to chronology. The span of astronomical time has been increasing dramatically even in the face of time-collapsing explosive events that reduce drastically the constraints upon time as a factor in change. Great stellar bodies exhibit rotations and motions that accomplish in hours phenomena that would on a gradual timescale be accorded millions or billions of years. It appears that one has to work with a paradox: even as one studies a Universe that changes over billions of years, one studies local events where changes are measured in microseconds. Consequence, which is the last hope of causality, is often strained in the straddling of time.

When the Solar System comes to be viewed in the light of newly discovered universal transactions, the idea necessarily arises that it has developed under time-collapsing conditions. Time measures - radiometric, geological and biological - that have been painstakingly manufactured to give billions of years of longevity to the system - must submit to a review of their credibility.

The need to generate a new chronometry is enhanced by current reassessments of legends and knowledge that ancient and prehistoric human beings possessed. The authors would not have ventured upon this reconstruction of the recent history of the Solar System were it not for the fossilized voices whose shouts about their catastrophic early world and sky sound louder even today than the shout heard in contemporary science about the exploding Universe. Those anthropologists, archaeologists, and scholars of ancient humanity who believe that these shouts must have been mere whispers confront the same impasse ideologically as those scholars who overlook the larger meanings of explosive cosmogony today. What the ancients said, and did not say, about the world are to be taken into account. Both their concepts of time and their visions of events deserve consideration.

This consideration and the others advanced before direct this monograph towards resolving the cosmogony of the Solar System into a model of a Solaria Binaria, the last stages of whose quick and violent quantavolution have been witnessed by human eyes. The model stands as plaintiff, confronting the model of uniformitarian evolution as adversary. Although a note on method is appended to the present work, it may be well to stress in the beginning that a prerequisite of scientificity is the ability to suspend judgment on a case being tried. This is especially painful when one is expert on the matter at issue. Even so, a scientist who cannot suspend judgment must be deemed as incompetent as the judge who cannot suspend judgment while hearing a case in a court of law.

As so often happens, what interests the public coincides with what interests scientists. Impelled by an intuition that is common to both the multitude of persons and the body of scholars, the human mind today is moving into an area "where the action is". For perhaps no more exciting and important a set of problems is to be found anywhere in the realms of science and scholarship.

PART ONE:

ORIGIN AND DEVELOPMENT OF THE BINARY SYSTEM

CHAPTER ONE

THE SOLAR SYSTEM AS A BINARY

Contrary to the hypothesis that the Solar System was born as and has evolved as a single star system, it is here claimed that the Solar System was and is a binary system. The binary system was formed when the primitive Sun fissioned. Several planets were generated in the neck of the fissioning pair and co-revolved about the Sun synchronously with the companion (see Figure 1).

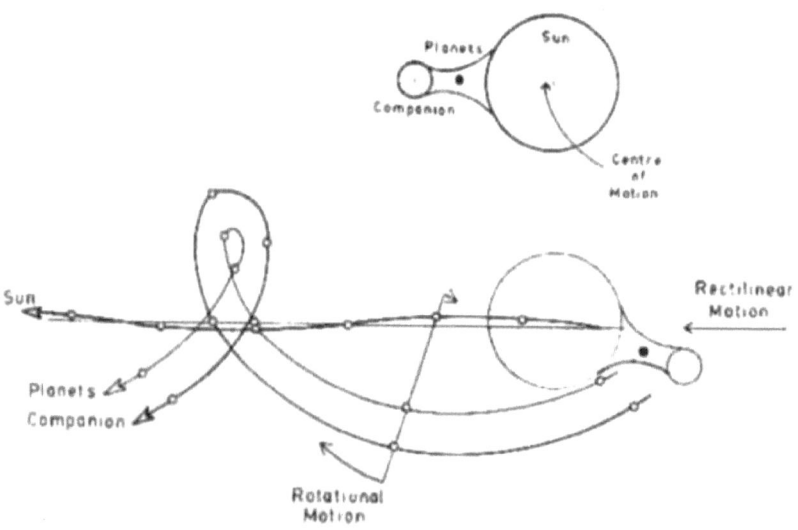

Figure 1
Dumb-bell Motion of Solaria Binaria

The binary system rotates like a lopsided dumb-bell as it moves through galactic space. The Sun orbits about the planets and the companion as they also orbit about the Sun. To be precise, all bodies in the system orbit about its center of motion with the same period.

The remaining planets were generated, one or more at a time, in several episodes, as the companion became unstable because of a changing galactic environment which we will discuss in Chapter Three.

Jupiter can be taken to be the remnant binary partner.[1] This => *quantavolutionary*[2] conception of a rapidly developed solar binary system is consonant with observations of nearby star systems. To seventeen light-years, or about one hundred million times the Earth-Sun distance of 150 => *gigameters*, there are forty-five star systems consisting of sixty stars and seven dark => *unseen bodies*. Among these are many => *physical binary systems*.

Sixty-one percent of the sixty nearest stars are components of a double or triple star system. Inasmuch as we cannot judge the organization of distant star systems, this statistic may or may not characterize the starry Universe. Even within our sample of sixty nearby stars, the star density and the binary frequency drop with increasing distance (van de Kamp 1971, p109), a suspicious fact.

Nothing that we know of the Sun is exclusively a property of a single star system or would be surprising if found in a => *double star system*.

On the average the => *principals* in a physical binary system are separated by approximately 18 => *astronomical units*. At one extreme, separations of up to twelve thousand astronomical units are deduced; at the other, the stars orbit one another with their surfaces in contact (see Technical note D, p. 223ff).

We see Solaria Binaria as a double star system evolving from the close extreme to a system showing increasing separation of the principals with time.

The typical => *visual binary system* that has been analyzed contains principals whose separations, periods, total masses, and orbital shapes are not markedly different from the Sun coupled with any one of the major planets of the present Solar system (Note D). The present Solar System differs from other visual binaries only when the => *luminosity* and mass ratios of the principals are considered. The observed features of visual

---

[1] We acknowledge the conceivability of a recent theory that a large remote planet or a dim distant companion of the Sun seems to be disturbing the planetary system (van de Kamp, 1961, 1971; Brady; Harrison, 1977) and might be a remnant binary partner in addition to Jupiter.

[2] See ahead to glossary.

binary systems are not an inconsistent final state for a physical binary system evolving in the manner that will be proposed here for Solaria Binaria.

The present mass ratio between the Sun and its planets would seem inconsistent with observed binary systems were it not for the fact that these latter are all visually observed and do not exclude the potential presence of binaries where the minor principal is undetectable presently by any observation. Further, as we shall show in Chapter Four the brightness of the Sun and its companion(s) was markedly different in the binary phase than in the present system.

The currently accepted cosmogony of the Sun and the planets is dominated by concepts of gravitation, great stretches of time, and the stability of stellar and Solar System motions. In this cosmogony one looks backward and forward in time, confident that the world has been and will be found in place under known conditions. One assumes the order of things in accord with a three-hundred-year-old theory backed up by centuries of systematic observations. Occasionally, but nowadays with increasing frequency, new scientific discoveries are "surprising" or anomalous, within the frame of the cosmogony. For instance, devastation has been wide-spread both on the Earth and on the other planets whose surface details are visible. Because theories had not predicted such instability, these disruptive events are insistently termed episodic and localized, and relegated to remote times. As will be shown, the prevailing cosmogony of science cannot cope with increasing numbers of surprising and anomalous observations. Sooner or later an alternative cosmogonical theory is invited. The mutating evidence suggests that a cosmogony can be constructed which does not require a long time to evolve our habitable world, within which major readjustments of the planetary orbits and environments are possible, and which redefines the set of forces that bring about change (see Technical Note C, p. 217ff)

We began with the theory that the Solar System originated as a binary star system and has evolved to the present as such. In the course of elaborating this theory, we shall have to develop and use new tools of analysis - a general concept of electricity (see Technical Note B, p. 212ff); new ways of viewing the origins of the atmosphere, lithosphere, and biosphere; an unusual form of legendary and historical inquiry (see Technical Note A, p. 202ff); and revised measures of time for the process.

Accepting the notion that the Solar System may be presently at the end of a long binary trail leads to a theory that the Sun is electrical. This fundamental idea is the topic of the next chapter.

CHAPTER TWO

THE SOLAR SYSTEM AS ELECTRICAL

The Sun, as star, radiates energy into the space surrounding it. Stars can be conceived to have originated from electrical cavities in the structure of space. Space, to our mind, is an infinite electrical medium. It is electrical in that it is everywhere occupied by a charge, which, when it moves, assumes the character of electrons, that is, "negative" charge (see Note B). The movement energizes and carries material into the cavities which become and are the stars.

Such electrical cavities or stars are observed in the millions, and inferred in the billions, in a fairly random distribution about the Sun. They form a lagoon of stars that is called the Galaxy, through which the Sun moves in a manner, and with consequences, to be described in the next chapter. Materially, a star is an agglomeration of all that has accompanied the inflow of electrical charges from surrounding space. The cosmic dust which astronomers see throughout the galaxies is matter yet to be forced into stellar cavities, or matter that has been expelled after a star dies. This dust is detected in greatest amounts in the vicinity of the most highly active stars[3].

Once in the cavity, the material cannot readily escape; it acquires

---

[3] To be considered is whether this may result from the dust in near stars being more observable.

increasing density because of electro-chemical binding and electrical accumulation. A cavity or star is increasingly charged but during its lifetime it cannot be more charged than the medium around it.[4] The Sun is highly charged, as some scientists have lately concluded (Bailey, 1960).

The life history of any new star may normally proceed as its cavity acquires first matter, and then charges continuously until its charge density reaches equilibrium with the surrounding medium, which is to say that the cavity has then been filled. Thereupon the star releases or mixes its material with the medium until it no longer possesses distinction as a body. This "normal" procedure is conditional upon the star's transacting with the space around it in a uniform manner. The majority of stars seem to transact quietly with their surrounding space, whether they are small red stars, or giant red stars. They end their existences as they lived, quietly, passing their accumulated material into the medium of space, eventually becoming indistinguishable from the medium itself.

However, the fact that the star is in motion within the galactic medium poses an occasional problem. It may journey into regions of the Galaxy which present it with greater or lesser electrical differences than it has been used to. Then quantavolution occurs. The star becomes one of the types to which astronomers pay the most attention - the variable stars, the highly luminous stars, the binary stars, the exploding stars.

It was in one such adventure in space that the original Super Sun lost its steady state, fissioned, and became Solaria Binaria. The system then consisted of a number of bodies, acting first as small "suns" with a primary partner, as is to be related in Chapter Four.

In recent times, according to the central theme of this book, this Solaria Binaria encountered a galactic region whose characteristics rendered the lesser stellar partner of the system unstable. In a series of quick changes the binary was transformed into today's Solar System.

Bruce (1944, p. 9) sees the process of stellar evolution as a cyclic build up of an electrically charged atmosphere above the star. As we see it, galactic potentials will determine the nature of the "surface" presented to the outside observer. As the star journeys through galactic space, its surface nature changes in response to differences in galactic potential. A change in the local galactic environment can lead to an instability which results in catastrophic electrical redistribution of the whole stellar atmosphere and

---

[4] The consequences of the temporary overcharging are described later when we consider stellar novae (Chapter Thirteen).

sometimes of material found well beneath the star's surface layer.[5] In short, the star becomes a nova.

In his cosmogony Bruce argues that binary stars form by division of an original stellar nucleus. When the star becomes a nova, the returning nova discharge, transacting electrically with the normal outward flow of => *stellar wind* off the star, induces the outbursting star to rotate. A possible reverse jet blast from the explosion might also cause the rotation to occur. Stars then, should have maximum rotation during the nova outburst. Fission of the star into a binary would then logically happen most frequently by rotational fission (Kopal, 1938, p. 657) immediately after a nova outburst. Close-binary pairs should be found among the post-nova stars (Clark *et al.*, 1975, pp. 674-6; Cowley *et al.*, 1975, p. 413).

The Solar System is probably the descendant of a Super Sun, a body containing at least eleven percent more material than the existing Sun, which became electrically unstable and underwent a nova explosion.

When the Super Sun erupted as a nova it divided into a close binary pair, whose primary became our present Sun; and its companion was a body about ten percent the size of the Sun (see Lyttleton, 1953, pp.137ff),[6] henceforth to be called Super Uranus. Enveloping the binary was a cloud of solar material constituting at least one percent of the Sun's material. Also created in the fission were the seeds which grew into the so-called "inner or terrestrial planets", probably Mars, the Earth, Mercury, and one that will be called Apollo. Apollo's fate is discussed in Chapter Fifteen.

Turning our attention to the Sun itself, we observe an opaque layer called the photosphere. This layer is regarded ordinarily as the Sun's surface. Above the photosphere lies the transparent solar atmosphere, which is difficult to observe. First comes the => *chromosphere* and then the corona. Perhaps the key to star behavior is the distinction between the photosphere and chromosphere. Each is examined and known by means

---

[5] See Bruce (1966b) for a discussion which compares a lightning discharges to the light curve for Nova Herculis 1934. Bruce (1944) mentions a discharge of the order of 10^{20} coulombs in the nova outburst. We see this atmospheric discharge as an electrical readjustment required after the star has responded to its changed environment.

[6] Lyttleton (1938) has argued that rotational fission cannot result in the formation of a stable binary system, but his arguments are probably invalid if the bodies at fission are highly charged (and of the same sign) but in different amounts (Note C). In this instance, immediate electrical transaction between the stars may allow non-collisional orbits to be stable, where they otherwise would not. Later criticism and support are well summarized by Batten (1973b). The arguments there about the stability of binary orbits over long times are in question because of the work of Bass. Likewise, the claim that fission cannot occur because stellar cores cannot remain uncoupled from stellar envelopes once rotational distortion becomes appreciable is also in question if the process producing the rotation begins in the envelope rather than in the core.

of spectroscopy, that is by observing and measuring its spectrum of => *radiation*.

The spectrum of the photosphere shows radiation produced when the atoms, => *ions*, and electrons of the photosphere collide, and therefore the spectrum reflects the state of atomic collisions there. The light is emitted during the collisions. It appears that the photosphere is a region of => *plasma* and atoms where the motion of the material is chaotic, randomized. Collisions occur after short journeys, after short mean free paths of electrical accumulation. The electrical field is small. A high kinetic energy of collision is registered in the temperature of several thousands of degrees. Energy is transmitted with some, but not great, amounts of conversion of energy into internal atomic structures (excitation).

By contrast, the spectrum of the chromosphere represents the release of the internal energy of excited atoms and ions. Light is emitted not so much at the moment of collision among atoms, but it is cast off by rapidly accelerating atoms moving to and from collisions, that is, between rather than during collisions. The chromosphere is a region of directed, vertically moving electrons descending into the photosphere, and atoms and ions escaping into the corona and the => *solar wind*. The mean free path is long, not short. The electrical field is large, not small as in the photosphere.

The photosphere, thus, is a region where the transmission of energy is observed. The chromosphere is a region where the => *transmutation* of energy is what is observed. The temperature "measurements" of the two regions are not helpful in understanding the dynamics, because in one case, temperature is "low" where short paths lead to frequent collisions, and in the other, temperature is high because of infrequent long-path collisions. What is important is the contribution of each region to the electrical system of the Sun.

The photosphere glows brightly with a silver color (Menzel, 1959, p. 24). Blemishing this visible face of the Sun are dark, slightly cooler regions called sunspots, the average spot lasts less than a day (Abell, 1975, p.527). Viewed by telescope, the whole photosphere, except where sunspots obscure it, shows a granular appearance. These => *granules* are bright patches, hot tufts of gas that live for only a few minutes (Juergens, 1979b, p.36).

The photosphere and the behavior of the solar atmosphere which lies above it can best be explained using a model based upon electrical processes. Bruce (1944, p.6), and later Juergens (1972, pp9ff) and Crew (1974, p.539) have shown that photosphere granules have the properties of a large number of parallel electrical arcs. Further, Juergens maintains that highly energetic electrons are transmitted from the Galaxy down through the solar atmosphere to the photosphere.

As in the Earth's atmosphere, the gas density and pressure in the solar atmosphere decrease with height above the photosphere. Where the atmospheric pressure falls to a value equal to one percent of the atmospheric pressure measured at the Earth's surface, collisions between gas atoms can no longer dominate the exchange of energy between the atoms. Instead it is the electrical processes that govern the energy exchanges in the solar gas. We see this transition as the hot chromosphere. The bladed or spiculed structure of the chromosphere consists of jets of gas moving upwards at about 30 kilometers per second. These spicules rise some 5000 to 20 000 kilometers above the photosphere (Abell, 1975, pp531ff) .[7]

Instabilities in the arc discharges lead to a build-up of charged regions in the solar atmosphere. These eventually produce electrical breakdown; sudden discharges occur, causing bright => *faculae*[8] and the temporary extinction of some photosphere arcs. The result is a sunspot (Bruce, 1944, p6).

The upper atmosphere of the Sun is the apparently intensely hot corona.[9] The gas atoms of the corona have been stripped of several electrons[10] by collisions with in flowing energetic cosmic electrons. The removed electrons are drawn towards the Sun so other ions can flow outwards into the corona allowing the coronal ions to recede into the solar wind. The spectrum of the lower corona shows the atoms stripped of several electrons emitting light between collisions, and the emission from the energetic electrons during collision.

The corona seems to be constantly ejecting its contents into space as the solar wind. The fraction of the solar output represented by the solar wind is about one-millionth. Haymes states that the whole corona is lost

---

[7] Juergens (1979b) believes the spicule is a fountain pumping electrons from the solar surface high into the corona. If he is correct, the upward motions detected spectroscopically in the spicules are produced by atoms bombarded by the electron flow. The electrons supplied by the spicules are necessary to allow ions to travel away from the solar surface.(See also Milton, 1979.)

[8] A facula (Lat : "torch") is a bright region seen best near the limb of the Sun where the underlying photosphere appears less bright.

[9] The temperature deduced from the spectrum is millions => *Kelvin*.

[10] Specifically, atoms heavier than helium which have lost several electrons are detected. In the corona, hydrogen and helium are present too, but cannot be detected since they have lost all of their electrons.

and replaced in about one day.[11]

Some of this material flows past the Earth's orbit as a cloud of energetic protons and helium nuclei, accompanied by electrons, known as the solar wind. In every second 100 million solar ions arrive above each square centimeter of the Earth's atmosphere.

The more luminous the star, the faster its stellar wind carries away mass, and, in general, the more rapidly the gases flow away from the star. Stellar wind flows of 10^{-10} to 10^{-5} Sun masses per year have been inferred with measured velocities from 550 to 3800 kilometers per second respectively (Lamers *et al.*, Table 1, p. 328).

Sudden explosive eruptions, called flares, occur above the solar surface. Energy in the form of light, atoms, and ions, is accelerated away from the Sun. The energy in a single flare could supply the Earth's population with electrical power for millions of years. A large flare releases in an instant about one-fortieth of the continuous solar output.

Flares start near sunspots, with associated faculae, and develop over hours. They move as if driven by an electrical potential difference between the Sun's surface and the higher atmosphere (Zirin, pp. 479ff, Obayashi, pp. 224ff). Once accelerated, the flare gases escape the Sun and modify the solar wind significantly. The cause of flares is baffling to conventional theories, which underplay electrical forces in cosmic processes. Most flare models involve some kind of magnetic driver to blow the gases from the Sun with great force (Babcock, p.420, p.422-4). The presence of magnetism implies an electric source. As we shall show in Chapter Six, the Sun once had an electrical connection to its companion, within which energy was released that created and sustained life within the binary system. Today's flares represent an undirected remnant of the inter-companion arc of yesteryear.

The solar wind consists of coronal gases which have been boiled away from the hot solar atmospheric discharges. It conducts the Sun's electrical transaction with the Galaxy. It is the Sun's connection to the Galaxy. The electron-deficient atoms (ions), by escaping from the Solar System, increase the negative charge on the Sun. This brings the Sun towards => *galactic neutral* and thus, in time, would end the Sun's life as star.

It follows that in the past, when the Sun was less negatively charged,

---

[11] Replacement of the corona in one day produces a loss of about 10^{-10} Sun's mass each year. Haymes' estimate for the loss of solar corona is much higher than the loss expected using measurements of the solar wind flux. One such solar wind measurement cited by Marti *et al.* would produce a corona loss which is 1/10 000 the value in Haymes.

more current flowed from the Sun to the Galaxy. Thus the present flow of solar wind is less than the flow in ages past when the Sun was more out of equilibrium than it is now.

The Solar wind varies with the ongoing "evolution" and "quantavolution" of the Sun.

In the past the solar wind flow was very complex because we believe that the Sun was a binary star and its companion, Super Uranus, was not in electrical equilibrium with it. The system eventually approached => *internal neutrality* because a large solar wind, electrically driven, flowed directly between the two principals.

In this connection we may explain the origin of the heavier elements in the Solar System. They were not built up from primordial hydrogen and helium, which show up so prominently in spectroscopic observation, but rather represent an accumulation in a period measurable in thousands of years of the fragments of heavy materials scattered initially near the Sun, near its binary partner, and along the electrified axis between the two (see ahead to Figure 7).

The theory that heavier elements are sparse in the interior of the Sun is probably incorrect. Spectroscopy cannot penetrate to beyond the photosphere; therefore it must show only a cloud of hydrogen admixed with metal and molecular vapors (Ross and Aller, Table 1, p. 1226) at low density.[12]

The mass of the Sun is calculated as a function of the orbital motion of the planets. Probably here, too, a methodological error is occurring that serves to produce the illusion of a light mass. Thus the model of the composition of the Sun depends upon the assumed structure of the solar interior and of the planets. If assumptions regarding both are incorrect, then the Sun's mass is probably incorrectly known.

Both incorrect theories - regarding the elements and mass - contribute to the major error of conventional Solar System theory, which is that the Sun is powered by thermonuclear processes, specifically the fusion of hydrogen atoms, in its interior.

Regarding the processes which power the Sun, most astronomers believe that there is an energy source deep in the solar interior obscured from view behind the opaque photosphere. If this belief is correct then the interior of the Sun must be hotter than the photosphere.

---

[12] Compared with the Earth's atmosphere, which at the surface has 1390 times the number of atoms per cubic centimeter as does the Sun's atmosphere at the photosphere.

Knowledge of the conditions within the Sun is inferred as the consequence of the physical forces *assumed* to be governing the stability of the Sun (Smith and Jacobs, pp. 223ff). It is usually inferred that near the center of the Sun the gas is sufficiently hot and dense enough to bring about => *nuclear fusion* on a large scale.

A thermonuclear Sun is an attractive theory since the Sun seems to be composed mainly of hydrogen. By compressing itself into a nuclear-powered core the Sun might radiate energy long enough to accommodate the gradual evolutionary processes believed necessary for the biological and geological developments that have occurred on the Earth.

However, thermonuclear fusion processes must dispose of large numbers of => *neutrinos*, and a vastly insufficient number of neutrinos have been detected on Earth in experiments specifically designed to capture the normally elusive solar neutrinos (Parker, p. 31). Before the nuclear Sun theory was presented, several mechanisms were proposed to explain the Sun's output of radiant energy.[13] All of these led to a radiant lifetime that was too short to satisfy the excessive time needs of the evolutionists.

Fatal, furthermore, to all theories of an internally powered Sun is the minimal temperature of the photosphere. How can the "surface" of the Sun remain cool when it is blanketed by hotter regions below and above whose temperatures reach millions of degrees (Parker, p. 28)? The usual answer is that the Sun's atmosphere is heated by turbulence within the Sun's outermost interior layers below the photosphere (Wright, p. 123). Somehow this process which, overleaping the photosphere, heats the Sun's atmosphere is supposedly divorced from the flow of radiant energy from the Sun's interior. Since such separation of processes is unknown elsewhere this explanation is unacceptable.[14]

Lastly, the observed turbulence (the granules) on the photosphere and

---

[13] Thus, the Sun, primordially hot, gives out heat as it cools; such a Sun has a life of thousands of years. Then Mayer, in 1848, supposed that the Sun is heated by in-falling meteorites. If they did the Sun would gain mass, affecting the size of planetary orbits. For his part, von Helmholtz, in 1854, showed that the Sun could radiate for tens of millions of years if it were contracting slowly. The reader is referred to the following sources for interesting and readable accounts of these mechanisms: Newcombe, Russell *et al.*, 1927; Rudeaux and de Vaucouleurs.

[14] Parker argues that a man (with a body temperature of 37° => *Celsius*) can rub two sticks together to ignite them (producing a fire at several hundred degrees Celsius). He adds that there is no limit to the temperature which can be obtained by so rubbing the sticks. What he fails to recognize is that if the sticks are continuously rubbed together generating heat by friction, they will conduct heat from the region of the friction. This heat will eventually reach the stick-holder's hands. Even if the stick-holder wears asbestos gloves, the wood, which is slowly becoming hotter, will eventually catch fire. On the Sun the photosphere must likewise heat up, unless it is somehow cooled by the warmer regions surrounding it. Such cooling is not spontaneous in nature.

its opacity are not compatible with the properties of hot gas of solar composition and condition (Juergens, 1979b, pp. 33ff). Since Bruce has shown the Sun outside the photosphere behaves like an electrical discharge, the theory, originally by Juergens, that the origin of the Sun's energy is external and electrical, is accepted here.

Consistent with the electrical phenomena of the Sun's atmosphere, we propose an external source of solar power. The Sun's light and heat output arises from the energy released by a flow of highly energetic electrons arriving from the Galaxy.[15] This electron current is enhanced by the flow of energetic solar wind protons away from the Sun.[16] The detected plasma a density near the Earth's orbit is 2 to 10 ions per cubic centimeter.[17] The ions flow outwards. Near Jupiter's orbit the Pioneer spacecraft measured no increase in the velocity of the solar ions over their velocity measured near the Earth.[18].

At the edge of the Solar System, escaping protons, accelerated to high energy by the drop in electrical potential between the Sun and the Galaxy, become galactic => *cosmic rays* and flow in all directions towards other stars. The protons expelled by other stars arrive in the Solar System as cosmic rays.[19] For energies above 100 GeV about six cosmic rays impinge upon

---

[15] The Sun's energy output is 4×10^{26} watts. If the arriving electrons have the minimum energy for cosmic rays not modulated by the Sun (see below, p. 33), which is about 100 gigaelectron volts (100 GeV), the inflowing current density at the Sun's photosphere would be 6.5×10^{-4} amperes per square meter. This value is a maximum; higher-energy electrons arriving lead to lower values for the electron current density.

[16] The flow of the solar wind particles is consistent with a potential barrier located at infinity (Lemaire and Scherer). Moving through the potential, the protons gain energy; as they flow away from the Sun and past the Earth's orbit the protons double their velocity, increasing from 150 kilometers per second in the corona to 320 kilometers per second at the Earth. The electrons in the solar wind show no drift velocity. The electron behavior is consistent with electrons being repelled by the distant Galaxy but also being repelled by a nearby Sun carrying an excess negative electrical charge, as was postulated much earlier by Bailey (1960).

[17] Zirlin remarks that spacecraft measurements of the solar wind plasma refer to protons, "but considerations of electrical neutrality require that the number of electrons per cubic centimeter equal the number of protons (although the velocities need not necessarily be the same)". Exact => *electric neutrality* cannot be assumed if the Sun is electrically powered from the outside, and thus we do not know the electron density in the solar wind unless it is measured.

[18] At the rate of solar wind flow, a sphere 100 AU in radius could be filled with plasma to 5 protons per cubic centimeter in about 10 000 years. However, moving at 300 km/s, a proton would travel about ten light years in this time, about 6300 times 100 AU. The material flow would be about 10^{17} tons (1/35 000 of an Earth).

[19] Conventionally, no origin other than "galactic" or "extragalactic" is ascribed to arriving cosmic rays not certainly identified with the Sun (Watson). The paucity of electrons in the cosmic ray flux is unconvincingly explained except by the notion of a star as an electron-deficient cavity in space.

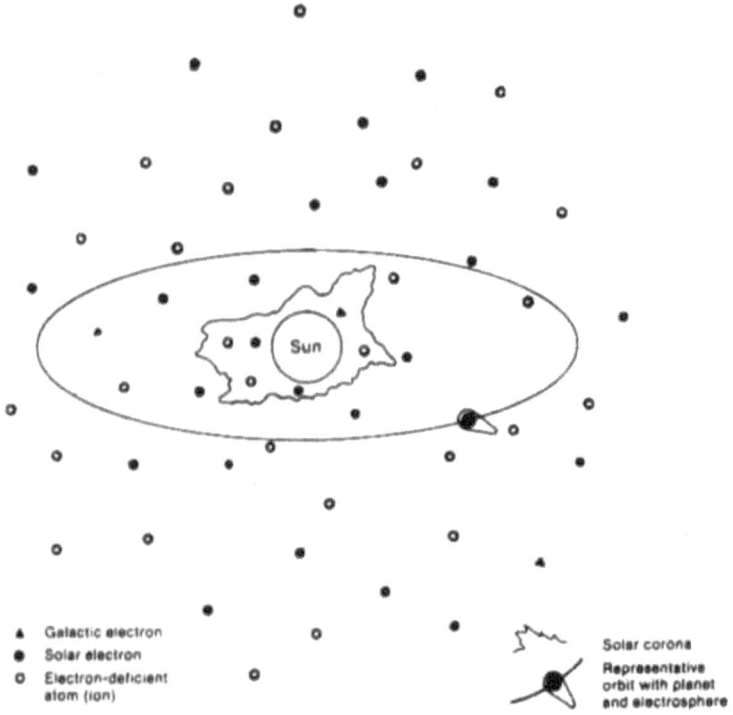

Figure 2
The Sun's Connection to the Galaxy

Outward-flowing solar wind ions carry an electric current between the negatively charged Sun and the more negatively charged galactic space that surrounds it. The solar wind flows through a "transactive matrix" (see Technical Note B) of solar electrons, which permeate the interplanetary space but do not flow through it as do the ions. Inward-flowing galactic electrons, travelling at velocities close to the *speed of light*, carry energy from the Galaxy to the solar "surface" where it is released and radiated as light and other electromagnetic waves, which constitute the solar luminosity.

each square meter of the Earth every second, but these few energetic particles carry inwards about one-twentieth of the energy flowing outwards with the solar wind at 1 AU.

That electron-deficient cosmic ray atoms continuously flow to Earth enhances the probability that the Earth is electrically charged. Juergens (1972) has argued that both the Earth and the Sun can have an excess (negative) charge.

At energies below 100 GeV the Sun somehow modulates the number of cosmic rays arriving in the inner Solar System (van Allen, p133). This presumably represents the maximum driving potential between the Sun and galactic space, with which it is transacting electrically. Cosmic rays with energy greatly in excess of 100 GeV would not be impeded meaningfully by the Sun's opposing driving potential.

Where the solar wind ends is yet to be determined. It was once believed the wind stopped inside Jupiter's orbit, later near Pluto, but today the wind is deemed to flow well beyond Pluto (Haymes, p237).

Somewhere the "galactic wind" meets the solar wind; there a boundary exists where the flow of incoming cosmic ray protons balances the outflowing solar wind protons. This is the edge of the Sun's discharge region, the limit of the Solar System.

To conclude, a star is born when an electric cavity forms in the charged medium of space, and matter rushes along with the charged space to fill the cavity. Then, after the cavity fills, the star dissipates into charged space, spilling out its matter simultaneously. No tombstone marks its demise; no derelicts travel forever through space. Indeed, existence is an attempt to achieve nothingness. Pockets of lesser negativity become existence by seeking to accumulate enough electric charge to emulate universal space, at which time they are capable of disappearing into nothingness.

CHAPTER THREE

THE SUN'S GALACTIC JOURNEY AND ABSOLUTE TIME

Conventionally viewed, the formation of a solar-type star and planets from a cloud of gases and cosmic dust takes on the order of several hundreds of millions of years. After accretion, an Earth-like planet supposedly takes another one or two thousand million years (1-2 gigayears or => *aeons*) to develop a stable lithosphere, which when formed allows the much slower evolution of a viable biosphere from the materials and energy available at the planetary surface (Oparin). To us, these processes seem too slow and rely too much upon random occurrences to be viable.

However, the processes forming stars and planets and leading to living things may proceed much more rapidly. Our cosmogony employs electrical cavities, charges and forces to accomplish change. These produce changes which are much more powerful and are highly selective.

Electrical force, as measurable by the repulsion between two electrons, compares with the apparent gravitational attraction of the same two electrons in the ratio of 10 36 to 1.[20] Conventional models of cosmic processes employ almost exclusively the trivially weak force termed gravity to produce and govern the Universe.

Electricity is a greater sculptor of change because it operates more

---

[20] Incidentally, the Universe, conventionally asserted to be held together by gravity, is said to be 10^{26} meters in radius; the atom, admittedly bound by electricity, has a radius of 10^{-10} meters. These radii are curiously in the ratio of 10^{36} to 1.

variably within a given cosmic setting. A simple lightning bolt can cause extensive surface damage, liberating megajoules of energy within a few meters of surviving observers. Only thousandths of a second are involved in the event. Yet, too, an undisturbed geological surface may be the setting for a large number of biological mutations provoked by a radiation storm of cosmic origin.

What "gravity" is supposed to accomplish in aeons, electricity could quickly accomplish before the eyes of the earthly observer. Driven by the powerful motivator, electricity, quantavolution becomes not only possible - but also essential. Furthermore an understanding of electricity's role provides a powerful new and *unified* explanation of most observable phenomena.

If the evidence cited in Chapter One has permitted us to proceed, viewing the developing Solar System as Solaria Binaria, and similarly, if in Chapter Two we end up viewing stars, and in particular, the Sun as an electric phenomenon, then we can hope to inquire about the time scale over which the Solar Binary developed. To be more specific, may we have a stellar binary which develops over a short interval through some of the most significant phases of the history of the Solar System ?

To tackle the problem of chronology we shall, as we have done before, look to the skies for the crucial clues. We must, in so doing, introduce a seemingly radical conception, one which we feel can be defended with the evidence to follow. We assert, in line with the past chapter, that stars take their properties less from the material which they contain and more from the electrical difference between the cavity, which creates the star, and the surrounding medium of electrified space (see => *space infra-charge*).

Translated into more common astronomical language, the luminosity of the star depends upon its galactic environment rather than upon the amount of material which it contains (see behind and to Technical Note B). The conventional notion that the more luminous the star, the more massive it is, was induced by Eddington from the analysis of a small sample of binary stars. As we interpret the same data, the more luminous the star, the more it transacts with its companion, and so the companion completes its orbit more rapidly (see Technical Note D). Unfortunately Eddington's Mass-Luminosity relationship is well established in astronomical formalism, so that today stars are assigned masses as soon as their luminosities are estimated.

There is a problem inherent in Eddington's method of massing the binaries. He calls upon "gravitational force" and nothing else to bring about motion within the binary system. The problem is compounded when luminosities are introduced as a way to measure mass in non-binary systems. Luminosity can only be known where the distance to the star can be measured. Star distances are computed using the annual parallax produced by viewing the displacement, as the Earth orbits the Sun, of any nearby star against the background of very distant stars. The parallax measurement involves measuring the minute angle at the apex of an isosceles triangle whose base is the diameter of the Earth's orbit about the Sun.[21] Parallax angles are very small; the closest star, Alpha Centauri, is only displaced through 1.52 => *arc seconds* over the year. This parallax, the largest, was not measured until 1839 (Baker, R. H., p. 317) Parallaxes are difficult to measure and they cannot be determined for stars farther from Earth than 652 => *light-years*. Such a small distance encompasses only one thousandth of the sphere of stars under close observation by astronomers. Thus the majority of reported star distances and luminosities are derived by theory rather than measurement. Of the twenty first-magnitude stars (the apparently brightest stars in the sky) only five are closer than 26 light years, the next five take us to 84 light-years; the next seven to 217 light years; and the last five to the measurement limit. In this sample are six supergiant stars; the parallax of one of these stars is only an estimate, two of the others are at the extreme limit, the last three are between 171 and 192 light-years distant. None of the most luminous supergiant stars are in this sample; thus all luminosities given for such stars are estimates! Even where parallax is measured, the measurement is rarely precise; uncertainties of 25% and larger are common, leading to luminosities which are most likely erroneous in the order of at least 56% (about half a magnitude unit). Near the measuring limit the possible deviations grow immensely, often exceeding considerably the number measured.

The famous => *Hertzsprung-Russell diagram*, the Rosetta Stone of modern astronomy, plots stellar luminosities against surface temperatures, determined from the star's spectrum. Since the spectrum is often difficult to classify, placement of the star on the diagram is not always easy (Baker, R. H., p. 342). To circumvent that difficulty astronomers now rely upon color indices in place of spectrum classes.[22] Such measurements are even more strongly theory dependent than the former in terms of their

---

[21] In practice, the parallax is half of the annual angular displacement of the star, and the base of the triangle, now right-angled, is one astronomical unit.

[22] The color index is determined by measuring the brightness of the star through two or more colored filters and comparing the intensities obtained with calculated laboratory profiles of intensity versus wave-lengths for various temperatures.

applicability to stellar emissions (see Wyse, p. 49), but they are more quantitatively formulated and therefore they lead to an unjustified sense of satisfaction with the computed result of the stellar condition. For our purposes they offer no help.

What we would say about the classification of stars is the following. In going from stars whose surface temperature appears to be high, to those which appear cooler, there is a gradation of the lines present in the stellar spectra. The hotter stars show absorption produced by helium atoms. As we look at progressively cooler stars the helium lines decline and abruptly hydrogen lines appear, increase in intensity, and slowly decline. As the hydrogen declines, the lines of the metals and metal ions increase in intensity through the solar type stars; they dominate in stars slightly cooler than the Sun, only to be surpassed in the coolest stars by band spectra produced by various simple molecules, notably hydrides and oxides. In some of the coolest stars compounds of carbon are prominent. Although astronomers may continue to seek a more precise classification for stars, we are content to employ the traditional spectral types for the present study.

Besides the Hertzsprung-Russell diagram that is used to classify the stars, astronomers have also divided the stars into populations according to their location within the Galaxy.

Some striking results were obtained:

1. The most luminous and apparently hottest stars are found within gaseous clouds containing much cosmic dust. These stars are confined in clumps to a thin plate that forms the equator of the Galaxy. Similar stars define the highly visible spiral arms seen in other galaxies.
2. Bright, cooler stars like Sirius are located near the equator of the Galaxy but are not confined to the galactic arms.
3. The disc of the Galaxy is populated with moderately hot stars (with 5000 to 8000 K surface temperatures); these stars resemble the Sun and populate the arms, the spaces between the arms, and make up part of the stars that occupy the central core of the Galaxy. These disc stars are the most numerous group of stars observed.
4. The disc of the Galaxy is enveloped in an ovoid shell of red giant stars whose spectra show fewer metals than stars of comparable type in the disc population. That these stars are mostly giant stars

is usually explained by claiming that the smaller stars in the population are not likely seen because of distance from the Earth. It is possible that the latter are absent. Most of what is known about these stars is from the study of giant stars within star clusters and intrinsically varying giant stars - where the star's luminosity varies in some characteristic way over an interval of days to months.

5. The Galaxy itself is embedded in a halo of cooler stars. Most of what is known of the galactic halo is deduced from a study of a few nearby small stars and 120 globular star clusters which surround the core of the Galaxy. One of these globular clusters, Messier 13 in the constellation of Hercules, has been described as a "celestial chrysanthemum" (Baker, R. H., p. 451). The number of stars in this cluster cannot be counted; but estimates around 500 000 are made. Averaging this number of stars over the volume of the cluster (not precisely known) it would seem as if the stars are about two light-years apart, much closer than the stars near the Sun. Some small halo stars are observed passing through the disc stars in the Sun's vicinity. Barnard's star is an example.

In summary:
- the most interactive stars and gas clouds form clumps which are the galactic arms
- around the arms is a disc of less interactive stars
- enveloping the disc are variously shaped ovoids and halos alleged to be progressively more "metal deficient" stars.

It has been proposed that the stars of the different populations of the Galaxy follow orbits about the galactic core which are characteristic of the population. Supposedly the arm stars have the most circular orbits; the disc stars follow slightly elliptical paths. Some are deemed to move inclined slightly to the galactic plane, like the asteroid orbits of the Solar System. The halo stars move in strongly elliptical orbits with random inclinations to the galactic arms, like the comet orbits of the Solar System. As they pass through the Sun's locality the halo stars betray their presence by large annual displacements compared to the disc stars.

All of the stars in the Galaxy are in motion. Since there is no standard of rest all we can detect is the motion of one star relative to another. Two streams of stars are observed moving past the Sun parallel to the Milky

Way (the arms of the Galaxy). The two streams move oppositely at a relative speed of 40 km/s, the outer stream moving towards Orion, the inner one to Scutum. These motions apparently reflect differences in the motion of consecutive galactic arm segments in the Galaxy. The stars in the Sun's "arm" we assume move with the Sun at 275 km/s [23] towards the constellation of Lyra near Cygnus, which is a motion away from the stars of Puppis.

Looking only at the net motion of stars close to the Sun we detect the drift of the Sun within its arm of the Galaxy. This analysis reveals a motion of 20 km/s towards the constellation of Hercules (away from the constellation of Canis Major)(Mihalas and Routly, p103).

Neither of the Sun's motions is precise but they should suffice for our purpose. The Sun's motion within its arm carries it four astronomical units per year. It takes nearly 22 500 years for the Sun to drift one light-year from its present position. But, when the galactic revolution motion is considered, the Sun is moving up to fourteen times as fast. In the extreme only 1107 years are required to displace the Sun one light-year, so in ten thousand years the Sun moves nine light-years, and in one million years it travels about 904 light years.

If our hypothesis is correct and the stars derive their properties from the space in which they are embedded, then a look at the stars presently in the Sun's wake will tell us how the Sun appeared in ages past. Unfortunately the path of the Sun over the last million years, within which we believe Solaria Binaria developed and collapsed, is not wholly within measured space. Luminosity assumptions need to be made during the first two-thirds of the binary's lifetime.

The Sun's total motion now is directed away from a point within the constellation of Right Carina (the solar antapex) at 8.4 hours Right Ascension and declination -62°.[24]

This antapex was determined by Strömberg using the radial velocities of globular star-clusters (Menzel *et al.*). In his sample, the Sun's drift and the Galaxy's revolution combine to produce a net motion of 286 km/s away from the antapex.

For star systems close to the Sun, adjacent stars are about 10.3 light-

---

[23] We choose this value from a list of several, spread between 167 ± 30 km/s and 300 ± 25 km/s, the values obtained using different samples of celestial objects (Mihalas and Routly). The choice can never be free of theoretical bias, nor of indeterminate bulk velocities possessed by the sample objects. Here, the choice is a compromise between accepted values for the galactic rotation (Menzel et al.) and the higher value derived from measurements within the Local Group of Galaxies (Mihalas and Routly).

[24] Negative declinations indicate coordinates south of the celestial equator.

years apart, each thus occupying a sphere containing 578 cubic light years of space (Allen, 1963, p237). Given such a low star density, a rather large volume must be examined around and along the Sun's wake to ensure that some stars are included. We have constructed, therefore, a cylinder thirteen and one-quarter light years in radius about the Sun's path. Moving for ten thousand years through this cylinder the Sun will "encounter" about 5000 cubic light-years of space. In such a volume there would reside about nine stars or star systems at the average local star density. Over the sixty-five light year swath through space covered by the *Gliese Star Catalogue* there are only fifteen star systems. It appears that along the Sun's path, the actual star density is only twenty seven percent of that expected. The Sun entered the region included within the Gliese catalogue about 74 000 years ago. Within that volume, our analytical sample of stars is reasonably complete . Beyond it, many of the stars located along the cylinder do not have published parallaxes and so they cannot be located in time; they cannot be used in the analysis.

The region of space which includes those stars which now occupy the space once passed through by the Sun on its galactic voyage is represented on a star map by a cone centered on the solar antapex [25].

The base of the cone in the present includes stars over one half of the sky. As time progresses backwards the frustum of the cone projected upon the sky diminishes in area (Figure 3). The frustum of the cone 3 500 years ago is a circle 76° in radius, encompassing stars from Orion's belt across the South Celestial Pole to the Scorpion's tail. Moving back twenty thousand years shortens the radius to 36°, thereby including the region from the feet of the Greater Dog to the Centaur's right foot. The area has only a 13.5° radius sixty thousand years ago; it shrinks to less than a 3° circle after three hundred thousand years.

Through recent time the Sun's trail is very close to a straight line projected towards the antapex. It is shown in Figure 4 and the stars included are listed in Table 1.

The stars occupying the space inhabited by the Sun through the current era (the Period of Solaria)[26] and during the time of the Late Quantavolutions, to be discussed in part Two of this book, are in this sample. Here, we find the nearest star system, the Alpha Centauri triple. The largest star is very similar to the Sun (Dole, p. 112).

---

[25] Because of galactic rotation the cone is bent slightly. Over one million years the path bends eastwards by a shade less than one degree , corresponding to a sideward displacement of 15 light years.
[26] See ahead to Table 6 (p. 124) for a summary of the periods during Solaria Binaria's lifetime.

Figure 3
Stars Around the Sun's Antapex

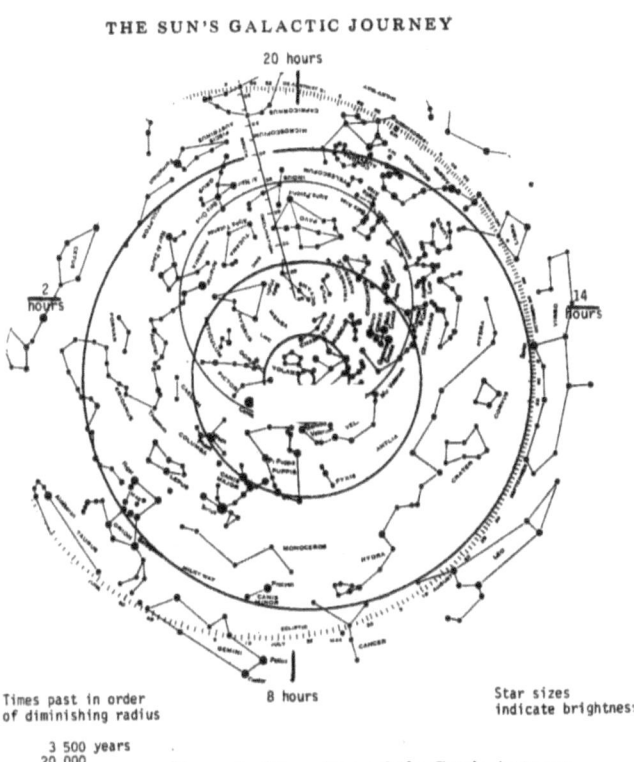

Figure 3. Stars Around the Sun's Antapex

The Sun's path traced backwards through the stars of the Galaxy passes through a cylinder of space whose axis stretches from the center of the Sun through the point on the celestial sphere with coordinates 8.4 hours of right ascension and -62° of declination. The edge of this cylinder, chosen to have a radius of 13.25 light years, is represented for different eras by the series of circles converging onto the solar antapex.

Its first companion is 23.5 astronomical units away moving along an elliptical orbit (Menzel *et al.*, p467). This star is slightly cooler and fainter than the Sun. The second companion is located almost two degrees away in the sky. It is over 550 times more distant than the separation of the closer pair. Frequent eruptions superpose bright emission lines on its otherwise faint class M spectrum. It is a flare star; its flaring might be associated with some intermittent transaction with the pair of distant companions. Unfortunately the α-Centauri triple is the only occupant within the space transited by the Sun during the series of quantavolutions preceding the historical period. It gives us no clue to an understanding of that space besides learning that solar-type stars can exist there.

Figure 4
Nearby Stars in the Solar Wake

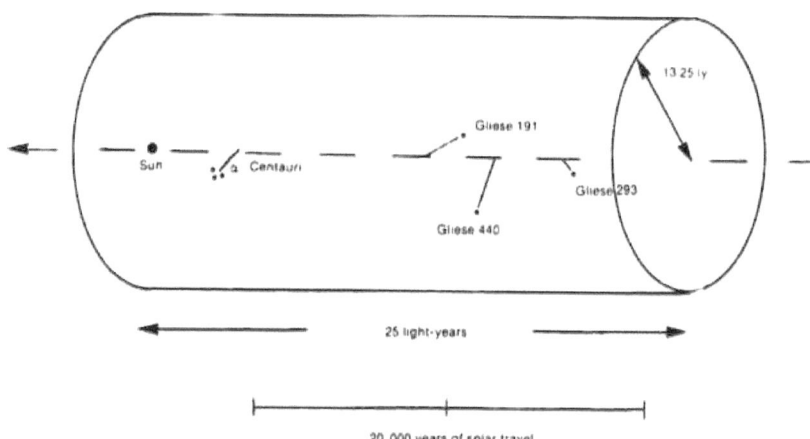

The sun's path through the space now occupied by the stars listed in Table 1. This space represents the region traversed by the Sun while it quantavoluted from Solaria Binaria into the Solar System we see today.

The three remaining stars are all low-transaction objects. This space we would suspect to hold a lower electric charge density than the space closer to the present. The closest of these three faint stars is located within the zone we believe was occupied by the Sun in the time before the eruptions began which eventually broke up Solaria Binaria. That instability of the recent past may well have been created as the Sun passed between the lower and higher regions of the transaction represented by these six nearby stars. The likelihood is that the Sun, late in the Period of Pangean Stability (Table 6), was less luminous than it is today.

Table 1
STARS BEHIND THE SUN (to 25 000 Years Ago)

Identification of Star	Distance from Sun (in ly)	Years in the Sun's wake (see Fig 3-2)
Alpha Centauri: Triple Star, main sequence components, dwarf "G", "K", and "M" stars; emission lines in the type "M" spectrum	4.3	4 860
Gliese 191: M0 main sequence dwarf star	13.0	14 750
Gliese 440: White dwarf start (class A)	16.1	18 200
Gliese 293: White dwarf start (class t-g)	19.2	21 700

Table 2

STARS BEHIND THE SUN (from 25 000 to 75 000 Years Ago)

Time (BP)	Star Name	Type
27 300	Gliese 257	M4 +
33 500	Gliese 341	M0
36 400	Alpha Mensae	G6
47 600	Gliese 269A	K2, Binary
53 500*	Gliese 333	M3
53 500	Gliese 375	M5 +
54 300*	Gliese 391	F3, Subgiant
64 700	Gliese 294A	F8, Triple
68 300	Gliese 298	M
73 800	Alpha Chamaeleonis	F5

Limiting magnitude + 18
* These stars are 25 ly apart, the Sun passes through space at their respective distances at the beginning and end of a 760 year interval.

Extending the Sun's line farther into the past to the limit of the Gliese catalogue (Table 2) we find no stars as luminous as the present Sun until we go back 54 000 years. Then along the path are positioned three stars that exceed the Sun in luminosity. The closest, an F3 subgiant, is five times more luminous; the second, the primary star in a triple system, is only 1.44 times brighter. Its two companions are very faint. The last of the three brighter stars exceeds the Sun's output eight-fold.

At the 75 000 year limit to Table 2 we reach the edge of the reasonably complete star sample. So far there are no conflicts with our theory. Stars of different spectral classes are well separated in space. In fact the cooler and hotter stars seem to be sorted: the class M stars tend to lie above the Sun's route while the class F and G stars are below it.[27]

If our calculated course is correct, the Sun's past behavior, as mirrored in the listed stars' present behavior, would show significant variation in luminosity over the tens of thousands of years represented here. Noteworthy, there are no highly luminous stars thus far along the Sun's trace.

Beyond 65 light-years, the magnitude limit of the available star catalogues containing measured parallaxes limits severely the completeness of the star sample. We can list no stars that are intrinsically fainter than today's Sun (Table 3). The catalogue from which the sample was taken covers only stars whose visual magnitude exceeds 6.25 (Becvar) whereas the Gliese catalogue includes known nearby stars above magnitude 18. Almost all of these stars show some distinguishing characteristic. The majority are binary, another has nebulous spectrum lines. These stars are positioned about the solar antapex in Figure 5. All could reflect plausible conditions for the early stages of Solaria Binaria's Period of Pangean Stability, and possibly also for the earlier Period of Radiant Genesis which followed the binary's creation.

At the limit of our proposed time (about one million years before present) using the Atlas of the Selected Areas (Vehrenberg) we count about 39 stars brighter than magnitude 12.5 in a target zone 40 by 40 arc-minutes adjacent to the Sun's antapex. Unfortunately no distances are given for the stars in this atlas.

---

[27] Given a small error in the solar motion (which is uncertain because the Sun's drift velocity, especially in the direction of the Galaxy's rotation, is variously reported with a twenty percent range), its path could be veering somewhat, either upwards or downwards relative to the path we have calculated. If so in this period the Sun might have become significantly brighter, or alternatively, remained much fainter than at present.

Figure 5
The Solar Antapex

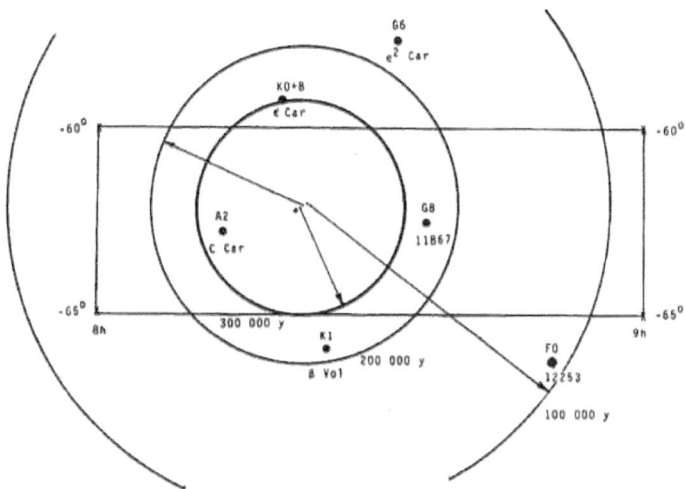

Map showing the brightest stars surrounding the solar antapex (see Table 3). The circles represent the described cylinder of space around the Sun at the ages shown. The successive radii are centered upon a slowly displacing point representing the solar antapex. The displacement, seen at this map-scale, occurs because the Sun rapidly orbits about the center of the Galaxy as it slowly moves through the arms of the Galaxy; its path therefore is a curved rather than a straight line.

Table 2
STARS BEHIND THE SUN (over 75 000 years Past)

Time (BP) (in Thousands of years)	Distance (in ly)	Star Name	Spectral Type
124	112	b Volatis	K1
134	121	C Carinae	A2, binary
139	125	GC 12253	F0, nebulous lines
258	233	GC 11867	G8, binary (M=+ 1)
301	326	e Carinae	K0, B; Spectroscopic binary

Limiting magnitude + 6.5 The sample ends at the edge of measured space.

Since our calculated solar target shows no stars the deficiency of the present measurable sample is confirmed. Nevertheless we see that the last listed star, 300 000 years BP along the Sun's run, is a spectroscopic binary whose class B primary is orbited by a class K secondary; a system not unlike our view of the early Solaria Binaria.

In our analysis more distant stars cannot be located in time along the Sun's path. Yet we can place, although uncertainly, several bright blue supergiant stars at locations surrounding the antapex in all directions and at distances corresponding to times between one-half and three million years ago. Several of the stars are components in binary star systems.

Within or on the periphery of this highly transactive region of space, the original Super Sun may have parturitioned to give birth to Solaria Binaria.

Although proof is hardly forthcoming from this analysis, at least evidence disproving the hypothesis is absent. We are encouraged to retain the idea that the behavior of star systems depends, if only in part, upon the celestial charge level of the space through which they pass. It seems as if this electric charge is contained not only by material residing in the space (stars, atoms, and electrons) but also, in part, as a charge embedded in the space itself, what we shall call a space infra-charge. Literally, the space infra-charge means that a vacuum (empty space) contains normally unavailable electric charges (here electrons) which generate the structure of that space and affect the behavior and properties of all matter occupying the space.

CHAPTER FOUR

SUPER URANUS AND THE PRIMITIVE PLANETS

About one million years ago, our Sun, then a Super Sun, underwent a nova eruption because of a sudden or unendurable change in electrical conditions. Solaria Binaria was instantly born. The Sun fissioned and in a huge blossoming cloud there would have been found a diminished Sun. Within a concentration of gases from the old sun would occur an admixture of chunks of the old Sun's interior material (nucleus), including a body that became the binary partner, which we here call Super Uranus. Between the new Sun and Super Uranus lingered other fragments of the fission and great quantities of the material that were to be absorbed into the planets. This impressive electrical quantavolution occurred in a matter of hours. The separation of the two bodies increased rapidly.

In electrical and chemical terms, we begin to detail this quantavolution. The normal flow of electricity between a star (the cavity) and the surrounding space is inward as is shown in Figure 6. The original Super Sun was such a star transacting quietly with the electron-rich space around it. The Super Sun became unstable, as outlined in Chapter Three, when its galactic journey carried it into a less electron-rich region.[28] Here, the enrichment presumably was rapid and of great magnitude, producing a quantavolution. The resulting nova, which is an explosion of electrons that

---

[28] The effect would be to make the star's surface suddenly quite electron-rich. Under such conditions the => *cosmic pressure* cannot hold the star's material together. The result is an explosive expansion. We cannot dismiss the possibility that a galactic electron storm suddenly enveloped the Super Sun, charging its surface to instability.

forces (requires) a material accompaniment, created Solaria Binaria. The Sun for a short time was relatively too electron-rich. In an explosive expansion the binary was born, not just from the solar atmosphere but also from the refractory materials normally hidden within its interior.

Figure 6
Electron Flow from Surrounding Space into a Star-cavity

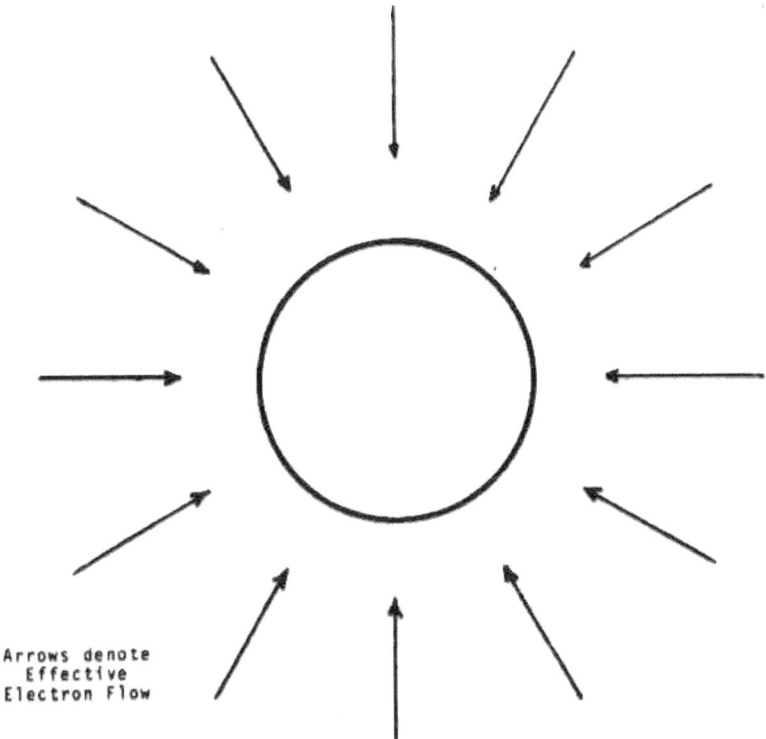

Arrows denote
Effective
Electron Flow

The Sun and the other stars represent electron deficient regions within the Galaxy. These regions, cavities as we call them, transact with the space around them gaining electrons during the lifetime of their central stars. When they become filled the stars they contain cease to exist.

The first state of Solaria Binaria is shown in Figure 7 below. The nova explosion had propelled what temporarily was excess charge away from the Sun. This of course would be illusory, for the Sun, by its continued existence, remained a region of relative electron deficiency. Thus, the initial dismemberment of the original Super Sun quickly halted: the expansion of the => *plenum* of material, now surrounding the Sun, ceased both because of the Sun's need for electrons and because the charged surrounding medium continued moving in upon the cavity. The boundary of the plenum shown above is actually a quantitative concept to denote the region where the outward pressure created by the charged Solaria Binaria is equal to the inward pressure normally produced by the Sun's galactic cosmic transaction.

At birth the electrical state of Solaria Binaria was radially layered. The system can best be described in terms of the local charge density of both the material and of the space into which the material was ejected in the eruption. The highest relative charge density existed at the perimeter of the plenum. Inwards this density decreased. The fragments ejected from the Sun, the debris forming the planets and Super Uranus, had progressively higher charge densities than the Sun, which had the least charge density in the system.

The Sun seeks its lost charge. The easiest way to get that charge is to launch into the plenum electron-deficient atoms (ions). The proximity of Super Uranus distorted greatly what otherwise would have been a radial flow of ions (as in the original transaction between the Super Sun and the Galaxy). A strong electrical connection coupled the Sun and Super Uranus; a lesser connection joined the Sun to the plenum, as shown in Figure 8 (see Technical Note E). This connection involved an inward flow of charge through the plenum. The charge flowed inwards either by direct transport of electrons or by indirect electron transport accomplished through the outward flow of electron deficient atoms (ions) (see Technical Note B).

The strongest electrical transaction occurred between the principals; accompanying this electrical flow, and highly influenced by it, was the transfer of material from one of the principals to the other. Elsewhere, close binary systems exist where the flow is from the companion to the primary (Cowley *et al.*, 1977, p471); more common is the flow from the primary to the companion (Mitton, p. 85, p. 100). The amount of flow and its direction would depend upon the distance between and the => *specific charge ratio* on the principals. We favor the flow of ions and gas from the Sun to Super Uranus. Since we often cannot resolve the principals into separate stars, designation of one as the primary and the other as companion is somewhat arbitrary. The choice usually is dictated by theory.

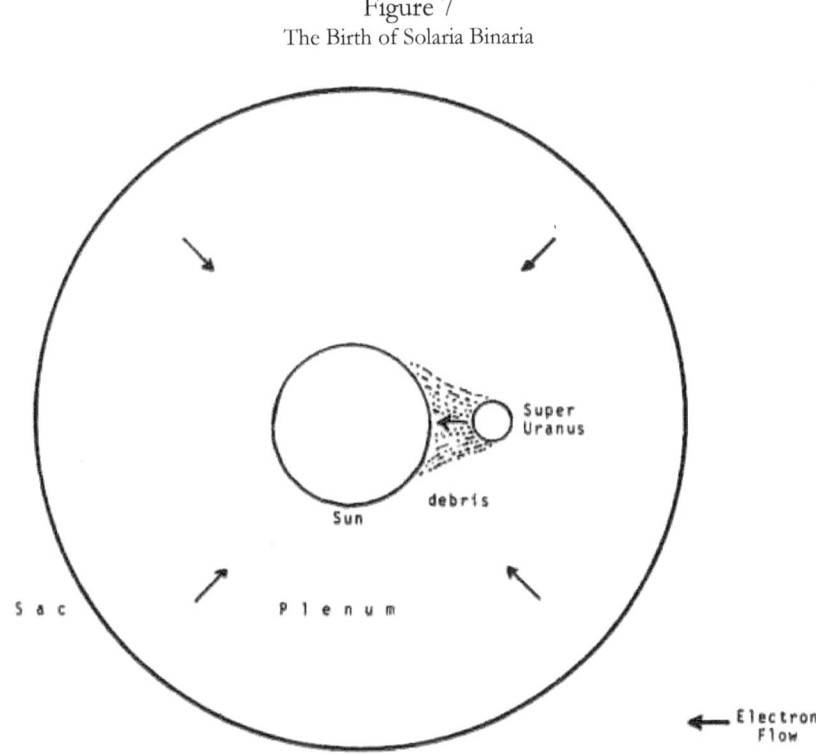

Figure 7
The Birth of Solaria Binaria

At its birth Solaria Binaria was embedded at the center of a plenum filling a sac of electron deficient matter. Electron flow into the sac from the Galaxy was augmented by electron redistribution within the plenum and among the components of the binary system.

Ionized gas atoms would be induced to flow between the principals. This flow of countermoving electrons and electron-deficient atoms would constitute a strong electrical current. As a consequence an intense magnetic field would be generated surrounding the current. This magnetic field would pinch the flowing ions producing a relatively narrow electrical flow channel (Zirin, p481). Collisions between neutral and electrified atoms would transfer the influence of the magnetic field (which affects only the electron-deficient atoms directly) to all of the gas between the principals; the result is a magnetic bottle (see Arp, pp. 213-5).

Figure 8
Material Flow Coupling the Sun, Super Uranus, and the Electrified Plenum.

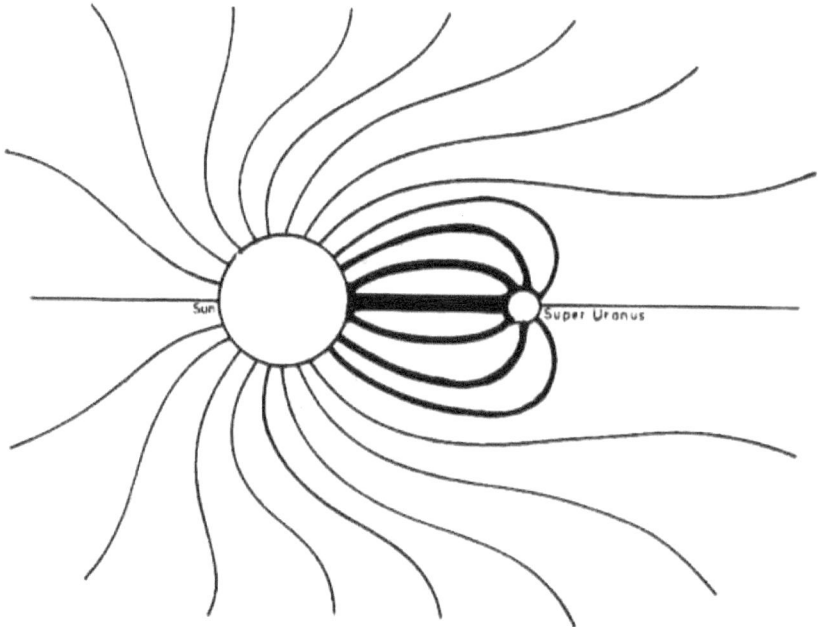

The creation of the Sun's companion, Super Uranus, greatly distorted the electrical flow between the electron deficient Sun and the Galaxy. The Sun's daughter, Super Uranus, like its parent, was short of electrons compared to galactic space outside the sac. The electrical flow coupling both the two stars and the stars with the Galaxy caused and directed a significant material exchange between the pair of stars

From the solar wind protons moving past the Earth, Juergens (1977c, p28) has calculated the current flowing away from the Sun in a sheet localized close to the ecliptic plane. If this same ion current was once flowing through the electrical channel, then the magnetic field generated was several thousand gauss in strength. Such a field would adequately constrain most of the gases producing a gaseous column or axis between the two stars. Material has been found along the interstellar axis in several binary systems (Batten, 1973a, p5).

Figure 9
Flow of Material Between the Sun and Super Uranus
under the Influence of a Self-generated Magnetic Field

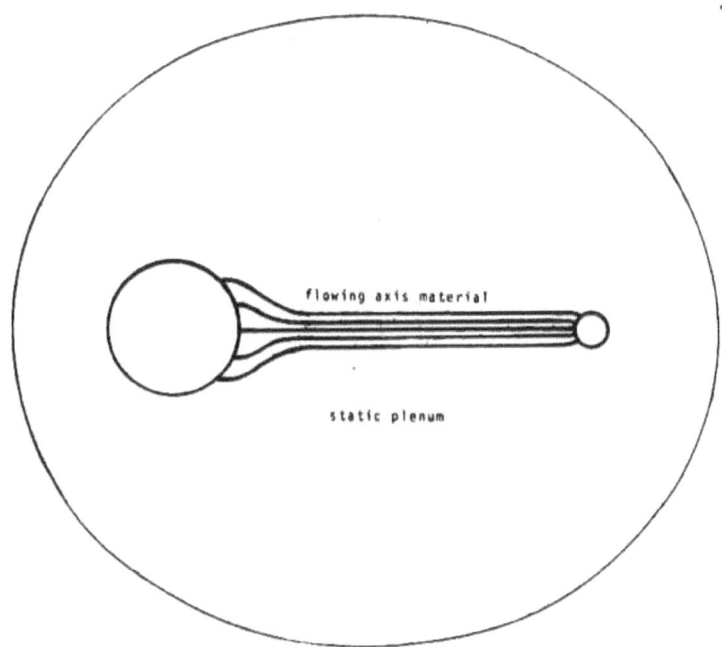

Electrically charged material flowing between the Sun and Super Uranus generated a strong magnetic field about the axis between the two stars. The effect of the magnetic field was to squeeze all material flow into a thin tube joining the stars. So constrained, the charged matter flow constituted a potent electric discharge, the arc, through the gases and matter of the plenum.

The absence of an appreciable interplanetary magnetic field despite the magnitude of the electric current represented by today's solar wind is understandable in terms of a planar current sheet model.

As shown in Figure 10, the solar wind sheet produces opposed toroidal (doughnut-shaped) magnetic regions above and below the planetary plane of motion. In the region between the toroids the magnetic fields generated by the radially diverging ions act so as to cancel out one another as in Figure 11. The vector sum of the magnetic intensity cancels between the parallel flowing ions but survives on their perimeter, leaving the postulated toroidal field. So, the regions above and below the Sun could be strongly magnetic, while interplanetary space so far explored lies outside of the toroidal field region, and has been shown to be almost devoid of magnetism. The existence of the magnetic toroid above and below the Sun may be responsible for the planarity of today's planetary region and the enhancement of the solar wind flow in that plane.

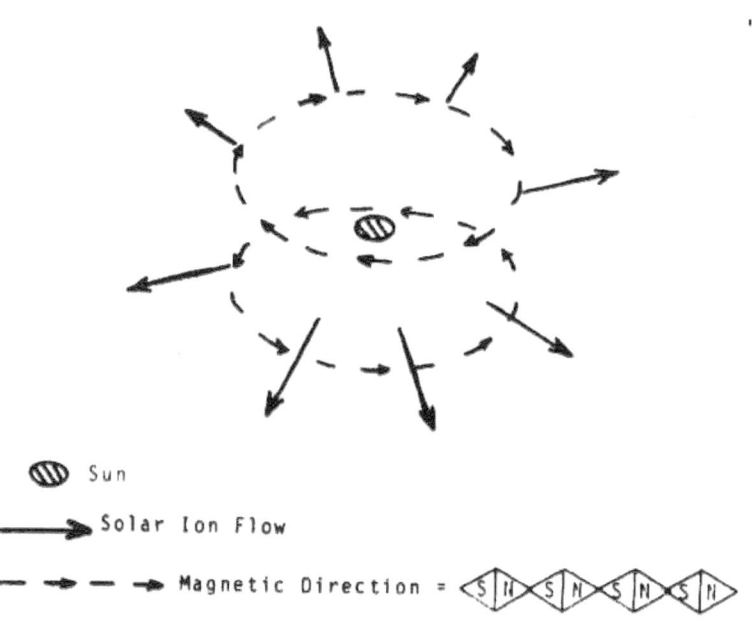

Figure 10
Magnetic Toroidal Field Produced by Solar Wind Current Sheet

Assuming that the solar wind is concentrated about the plane of the orbiting planets, the outward flow of ions from the Sun would represent a sheet of electric current. A significant magnetic field, curved upon itself to form a doughnut (a torus), would be generated by the existence of the solar current sheet. This toroidal magnetic field should be found in the space above and below the space occupied by the solar wind.

The Sun's rotation began consequent to the nova discharge creating Super Uranus. Super Uranus thereafter wheeled about the Sun in close orbit. The magnetic field produced by their electrical transaction was instrumental in locking the rotation of the Sun to the motion of Super Uranus about the Sun. Strongly coupled together the pair rotated looking like an ever expanding but otherwise rigid dumb-bell. The gases and the planets as they formed remained trapped along the gaseous electrified axis between the principals.

Figure 11
Magnetic Field Surrounding Several Flowing Ions

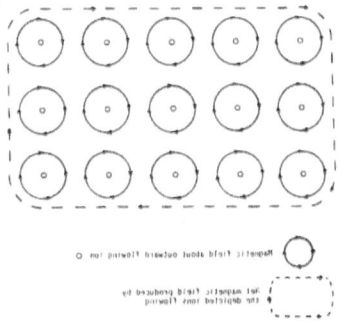

Each moving ion (or electron) comprises a unit of electrical current. It generates a magnetic field which appears in the plane perpendicular to its motion. When electrical charges flow radially, as does the ion wind from the Sun, only a tiny magnetic field is apparent in the region between the flowing ions because the magnetic effect of each ion is cancelled by that of its neighbors. A significant magnetic indication of the electrical flow is found only along the perimeter of the current sheet produced by the radial flow of the ions.

Plavec notes that the companion, if less massive than the Sun, can always be expected to rotate in synchronism with orbital motion. He states, also, that for all binary systems synchronism of rotation and revolution seem to occur for orbital periods shorter than ten days. For longer periods the synchronism falls except as postulated above.

Batten (1967, p. 36) notes that some semi-detached binary systems, particularly the Algol group, have primaries which rotate appreciably faster than would be expected for orbital synchronism.[29] We see these systems as a later stage of evolution of the binary. Solaria Binaria did not detach in this way until after the Saturnian period (see ahead, Chapter Fourteen).

The evolution of Solaria Binaria was such that the two principals were slowly driven apart, in part by the momentum of the flow of mass from one to the other and in part from increased repulsion caused by the growing level of electric charge in the whole system by the accumulation

---

[29] In cases of anomalous primary rotation, the anomaly is generally detected because the spectrum lines of the primary star are unusually bright. This line broadening could be, as well, evidence of electrical fields within the star's atmosphere (Stark effect).

of galactic electrons. All the while the angular momentum (spin) within the system was being transferred from the primary to its companion. At fission the Sun could have had over 80 percent of the angular momentum. The evolved binary (today's Solar System) left less than one percent of the angular momentum in the Sun. If matter was transferred mechanically from the heavier Sun to the lighter orbiting Super Uranus, the spin of the binary would decrease, but if the transferred matter is electrically driven, acceleration would be expected to accompany the transfer, thereby potentially increasing the spin of the binary. Even if no increase in spin occurs and even with a slight slowdown of spin, angular momentum is slowly lost by the Sun and gained by its companion and the primitive planets as the electric transfer continues.

The pulling apart of the principals was reflected in an increase in the binary's period of revolution. That is, Solaria Binaria wheeled more slowly about its center. There is a significant relation between the period of revolution of binaries and the observed "surface temperature" of the primary star. Certain stars called => *early-type* by astronomers tend to have companions with shorter periods (Russell *et al.*, 1927, pp.703ff).

In its earlier stages, Solaria Binaria would have looked to a distant observer as a close binary with an unseen companion. We imply that the Sun was an early-type star but not in the usual sense of the term star. Within the => *sac*, where the two stars and the Earth were located, the energy flow may always have been similar to what we observe now. However the outer parts of the sac were transacting intensively with the cosmos and thus were radiating so as to appear markedly hotter. The perimeter of Solaria Binaria, then, would have appeared to radiate as an early-type star and not like the Sun does now (see ahead to Figure 21). Its period of light variation, radiation emitted, and flow of mass would have attracted the attention of astronomers elsewhere to Solaria Binaria.

Some curious "age disparities" exist between principals of binary systems. In the Sirius star system, a young => *main sequence* star is orbited by a less massive old white dwarf star (see Kopal, 1938). The B-emission stars (hot, very rapidly rotating main sequence stars surrounded by a shell of gas) are often spectroscopic binaries whose companions orbit in about ten days. The companion is usually invisible and believed to be a highly => *evolved star* relative to the primary (Maraschi *et al.*). The highly evolved component admittedly often has so little mass that a nuclear synthetic evolution (see => *nucleosynthesis*) could not have aged it so rapidly (Kraft).

Both the age disparities and the size anomalies disappear if electrical evolution is considered. It is noteworthy that many of the interesting close-binary systems involve an unseen companion. The primaries in these systems range from very hot-type O-stars to very-cool-type M-stars. The sizes and masses within these star systems are inferred conventionally from the theory of evolution for the thermonuclear star (see => *thermonuclear fusion*). We do not agree with such an interpretation of this evidence.

We will not pursue the stages of early evolution of Solaria Binaria here (for that, see Part Two). The first aware men saw the skies in the => *Age of Urania* about thirteen thousand current years before the present (de Grazia, 1981). There were no humans capable of comprehending Solaria Binaria before it began to break up at the end of an Earth age that we shall be calling Pangea.

Super Uranus was first revealed to humans as a luminous object about twice the size of the Sun we observe today. The Earth was then located about two-thirds of the distance from the Sun to Super Uranus, because it was still electromagnetically bound to the axis between the stars. The objects found within the inner regions of galaxies seemingly orbit in this way - and probably for the same reason.

With such a configuration the Sun, if visible, would have been seen from the Earth's southern hemisphere only and would appear 2.5 times larger than Super Uranus, which in its turn was visible only from the northern hemisphere. The hemispheres referred to here are not those inscribed on the Earth-globes of today. They refer to the ancient references to the sky gods and their places. The Earth moved with its "north" locked towards Super Uranus (see ahead to Figure 18). No other major gaseous planet was in existence at this time. As the solid-wheel binary evolved, the Sun eventually was separated from Super Uranus by 105 gigameters (about 0.7 astronomical units). Before the next great quantavolution the primitive planets Mars, Earth, Apollo, and Mercury ended up between the two principals in the region between 61 and 96 gigameters from the Sun. At such separation this would bring the planets Mars and Mercury closer to Earth by factors of four and six respectively. Even so these planets would produce visible discs which were only about one twenty-seventh the size of today's Moon. If they could be seen (which we doubt) they would still be observed almost as points of light in the sky.[30]

The planets were originally debris from the Super Sun nova. They traveled out in the trail of Super Uranus, held in the electric and gaseous flow. They settled into their original positions rather than moving on

---

[30] The resolution of the eye is at best 20 arc-seconds; for night vision resolution is much worse than this (Greenberg, L. H.).

because they were electrically less negative than Super Uranus. They distributed themselves in their magnetic cage along the axis in accord with the principle of maximum mutual repulsion (elsewhere known as "the principle of least interaction action;" see Ovenden, 1974).

Several cosmogonies involve processes occurring within a binary star system. Gunn proposes that planets arise from the break-up of a rotationally unstable star, the same process by which he accounts for the formation of a binary star system from a single star. Lyttleton (1936, p. 559) visualizes a process by which planets form during an encounter of a star with two other stars; for such an encounter between three stars to be likely the stars must formerly be members of a bound system of stars, a triple star system. Bruce (1944, p13), like Gunn, sees the process of planet formation as a special case of fission of one star into a binary.

From the beginning Solaria Binaria was enveloped in a cloud of solar material (gases and solids). As the binary evolved this sac became extended along the lengthening axis from Sun to Super Uranus. Compressed by the magnetic field generated by the flowing electrified gases, a stable gaseous tube surrounded the planets; indeed these gases pervaded the entire planetary region, enveloping all of the planets in a single sac of gases. Within this dense gaseous sac, the contents of which the ancients called the aether,[31] and we will call the plenum, the planets could receive biologically necessary temperatures from the axial electrical discharge connecting the Sun with Super Uranus (de Grazia, 1981). If today's aircraft had existed then, they might have flown regularly among the planets.

The approximate size of the gaseous tube within which the Earth and the other planets moved was at most the diameter of the Sun, and at the least a significant fraction of the diameter of Super Uranus. This tube confined the plenum which allowed life to develop and thrive on all of the planets of Solaria Binaria.

---

[31] See Aristotle (Astronomy), where he argues that the outermost regions consist of an elementary kind of matter which is distinct from the other elementary substances (earth, air, fire and water). Also, in *Meteorology*, he notes that Anaxagoras thought that the upper regions were burning hot. Anaxagoras called the substance which prevails in those parts *Aether*. Aristotle adds that the ancients assumed that the *aether* is an eternal substance whose motion never ceases. It is like nothing else we know. There was controversy among the ancients as to whether the term aether (GK. *aither*) is derived from *aeithein*, " to run always", or from *aethein*, "to burn". Aristotle favors the former (Gershenson and Greenberg), although Anaxagoras and modern etymologists prefer the latter.

CHAPTER FIVE

THE SAC AND ITS PLENUM

The original Super Sun, prior to its nova, was accumulating electrons from the Galaxy consistent with the demands of the environment through which it was passing. As we have explained earlier, the Super Sun became too electro-negative and expelled material violently into its surrounding space. This material could not escape; its expulsion was opposed both by the post-nova Sun and by the Galaxy. It thus formed and filled a sac surrounding the newly created Solaria Binaria.

In the sac was the whole system of Solaria Binaria; the Sun, Super Uranus, the primitive planets, and the plenum (of gases and solids) of solar origin that nurtured the planets.

As the binary widens, the sac becomes conical in shape, narrowing from the size of the Sun at one end to about the size of Super Uranus at the other. A system of similar appearance has been postulated for the binary AM Herculis (Liller, p. 352). Wickramasinghe and Bessell describe gas flow patterns in X-ray- emitting binary systems. There, one may note a similarity in the shape of their pattern of maximum obscuration to the cone of gases proposed in this work.

Viewed from the outside the ancient plenum would have been opaque to light. Not so with the gas of the Earth's atmosphere today, which is eight kilometers thick if the atmosphere is considered as a column of gas of constant density.[32] This atmospheric layer is of trivial thickness compared to the radius of the Earth, yet its importance to the environment is unquestionable. Even this negligible atmospheric layer removes 18.4 per

---

[32] The actual atmosphere does not have a constant density throughout its volume. If condensed to constant density it would become an 8-km column of gas at the atmospheric density found presently at the bottom of the atmosphere.

cent of the incoming sunlight, mostly by diverting it from its original direction of travel.

Some of this scattered light returns to space, but most of it is redirected several times to produce the blue sky so familiar to us. Atmospheric scatter is enhanced near sunset when the incoming light traverses an atmospheric column tens of times longer than near noon. The setting Sun is notably fainter and its color redder because of the increased scatter. If the atmospheric column were as little as 1280 kilometers thick (at the present surface air density) all of the sunlight would be deflected from its incoming direction. Light would still be seen but only after scattering several times; no discernible source could be identified with the light. So it was in the days of Solaria Binaria. To be precise, if, in the last days of Super Uranus, this body were about thirty gigameters from Earth and if Super Uranus was then as bright per square centimeter of surface as today's Sun, it would not have been directly visible unless the gas density in the plenum was close to that deduced today for the Earth's atmosphere at an altitude of eighty kilometers. To see the more distant Sun this density would have to be decreased another fourfold.[33]

In the Age of Urania, Super Uranus was located about as far from the Sun as the orbit of the planet Venus today. This would provide the plenum with a volume of about 10^{20} cubic kilometers. If the plenum contained as much as one per cent of the atoms in the present Sun, the gas density would be several times that found at the base of the Earth's atmosphere today. Neither star would be seen directly, and only a dim diffused light could reach the planetary surfaces.

As the binary evolved, the plenum came to contain an increased electrical charge; it expanded, leaving less and less gas in the space between the principals. Thus it became gradually more transparent.

Astronomers see diluting plenum gases elsewhere in evolving binary systems. Batten (1973a, p. 10), discussing matter flow within binary systems, favors gas densities of the order of 10^{13} particles per cubic centimeter. Warner and Nather propose a much higher density for one system (U Geminorum-a dwarf nova system) where they postulate a gas disc with 6×10^{17} electrons per cubic centimeter. Unless all the gas is ionized, the neutral gas density would be higher than the calculated electron density. The gas densities that they mention are comparable to

---

[33] The retention of a more dense, thin atmospheric skin surrounding the Earth (and the other planets) would not affect the visibility of the binary components more adversely than does the Earth's atmosphere today.

those necessary to allow the early humans to discern the first celestial orbits.

In the earlier stages of Solaria Binaria the plenum was impenetrable to an outside observer; all detected radiation came from the surface layers of the cone-shaped sac, an area up to fifty-five times the surface of the Sun. The luminosity of the sac would arise from the transaction between inflowing galactic electrons and the gases on the perimeter of the sac.

The plenum, at formation, was electron-rich relative to the stars and the planetary nuclei centered within it. These latter electron-deficient bodies promptly initiated a transaction to obtain more electrons by expelling electron-deficient atoms into the volume of the plenum. The charge differences within the sac were modulated with time. In other words, the plenum was losing electrons from its perimeter to its center. In response, the size of the sac collapsed under cosmic pressure. In time this charge-redistribution might have diminished the volume of the sac by as much as tenfold, compressing the cone of gases into a cylinder or column of smaller diameter.

Running along the axis between the Sun and Super Uranus was an electrical discharge joining the two principals. Moving with this electrical flow was matter from the Sun that was bound for Super Uranus. Some of this matter would be intercepted by and incorporated into the primitive planets.

Induced by the electrical flow a magnetic field was generated which encircled the axis and radially pinched the gases. The pinch effect is self-limiting in that the more the current, the more the pinch. An infinite current in theory pinches the current carriers into an infinitesimal volume, extinguishing it (Blevin, 1964a, p. 214). Material would be extruded at both ends of the pinched flow by the pressure induced in the pinch.

This circular magnetic field, a magnetic tube, would induce randomly moving ions of the plenum to circulate along the field direction. The circulating motion of the ions eventually would be transferred by collision to the neutral gases. The result would be that in the outer regions flow would be dominated by revolution around the circumference of the tube. Everything here would eventually revolve uniformly. The innermost regions of the column were dominated by flow along the axis. Considerable transaction occurred at the junction of these two separately moving regions of the column, the central and the peripheral.

Some luminosity would arise from the transaction of electrons and ions deep within the magnetic tube. The ions electrically accelerated towards

Super Uranus were neutralized at some point along their trajectory. At neutralization X-rays were produced. Some of the ions would be neutralized upon collision within the magnetic tube, most upon reaching Super Uranus; but, because of the pinch phenomenon noted above, some ions would be extruded and neutralized near the perimeter of the sac behind Super Uranus. Despite the high gas density in the original plenum, X-ray emission would be observable from the outside. That such is the case elsewhere is indicated by Brennan.

As the plenum diluted with time (in a manner to be discussed in Chapter Eleven) the outside observer would see deeper and deeper into the system, and eventually all of the X-ray emission would come from the interface between the magnetic tube and the surface of Super Uranus. As in other binary systems, a partial eclipse of the main X-ray source would then be seen as the dumb-bell revolved (see Tananbaum and Hutchings for data on other binaries).

Matsuoka notes a positive correlation between X-ray and optical emission in binaries. Radio-emitting regions surround many binary systems (Wickramasinghe and Bessell). Spangler and his colleagues claim that radio emission from binary stars is noted for stars that are over-luminous. The radio emission is generated by electrons transacting with the magnetic field associated with the inter-star axis. That this emission is enhanced when a stronger transaction occurs between the stars causing the over-luminosity is understandable, using our model.

At the perimeter of the plenum, optical effects would show to an outside observer an apparent absorption shell associated with the hidden binary within. Like many of the close-binary systems, the stars of Solaria Binaria would not be resolvable in a distant telescope, but the binary nature of the system could be known because observable differences would be produced as the dumb-bell revolved.

Gas-containing binary systems as described here, and elsewhere (Batten, 1973b, pp. 157ff, pp. 176ff), represent the stake of Solaria Binaria at various epochs, and especially in its last days. As the binary system collapsed, the plenum thinned, allowing direct observation of light produced by sources inside the sac. The gas disc, theoretically implied to surround the stars of other binaries, is waning in the late translucent plenum. The gas streams detected flowing between certain binary components are present in Solaria Binaria along what we call the electrical arc. The gas clouds, whose absorption spectrum leads us to believe that they envelop entire binary systems, correspond to the perimeter of the early opaque plenum. As Solaria Binaria evolved, each of the classes of circumstellar matter noted by astronomers became observable in their turn.

Inferable from the above is the degree of visibility from the Earth's surface, or from any point of the planetary belt within the plenum. Overall there is a translucence. Objects near at hand might be distinguished, certainly after the half-way mark in the million-year history of Solaria Binaria was past. Sky bodies were indistinguishable from Earth.

With passing time, the level of light would increase. In the beginning, the light is scattered and the sky is a dim white. As the plenum thinned electrically, the sky bodies would emerge as diffuse reddish patches. During this process, the sky would brighten and become more blue. Thus, as they emerge, Super Uranus and the Sun brighten and whiten while the sky becomes darker and bluer.

At a time related to the changes soon to be discussed, around fourteen thousand years ago, the Earth is suddenly peopled by humans, and one may investigate whether any memories remain of the plenum. There seem to be several legendary themes that correlate with our deductions about visibility.

Seemingly, aboriginal legends describe the heavens as hard, heavy, marble-like and luminous. Earliest humans were seeing a vault, a dome.[34] Probably in retrospect, to the heaven was ascribed the human qualities of a robe or covering, and, by extension, part of an anthropomorphic god. Thus, the Romans saw Coelus, the Chinese T'ien, the Hindus Varuna, and the Greeks Ouranos. Vail (1905/1972) presents ample evidence that day and night were uncertain and that the heavens were continuously translucent. When Hindu myth says that "the World was dark and asleep until the Great => *Demiurge* appeared", we construe the word "dark" as non-bright relative to the sunlit sky that came later. Heaven and Earth were close together, were spouses, according to Greek and other legends. The global climate of the Earth in the plenum was wet; all is born from the insemination of the fecund Earth by the Sky, said some legends. There was so much moisture in the plenum that, although the ocean basins were not yet structured, the first proto-humans might confuse the waters of the

---

[34] Vail (1905) collected ancient expressions from diverse cultures testifying to perceptions of the heavens as "the Shining Whole", "the Brilliant All", the "firmament", "the vault", "Heaven the Concealer". Heaven was the Deity who came down crushingly on Earth, and the heavens are said to "roll away" and to open to discharge the Heavenly Hosts; great rivers are said to flow out of Heaven. In other places we read of the gods chopping and piercing holes in the celestial ceiling, of a Boreal Hole that is an "Island of Stars", a "star opening", "Mimer's Well". Heaven was perceived to become ever more impalpable and tenuous with time, so that not only the memory of it but also its names, adjectives and metaphors lost their strength of meaning.

firmament above with the earth-waters. In some legendary beginnings, a supreme deity had dispatched a diver to bring out Earth from the great primordial waters of chaos (Long, 1963).

The earliest condition was referred to as a chaos, not in the present sense of turbulent clouds, disorder, and disaster, but in the sense of lacking precise indicators of order, such as a cycle that would let time be measured. T'ien is the Chinese Heaven, universally present chaos without form. The gods who later give men time, such as Kronos, are specifically celebrated therefore (Plato).

Sky bodies were invisible. Legends of creation do not begin with a bright sky filled with beings, but speak of a time before this. When the first sky-body observations are reported, they are of falling bodies. The earliest fixed heavenly body in legend is not the Sun, the Moon, the planets, nor the stars, but Super Uranus, as will be described later on.

Nor was the radiant perimeter of the sac visible. It lay far beyond discernment as such, and was in any case practically indistinguishable from its luminescence. The electrical arc would have been visible directly only in its decaying days, being likewise sheathed from sight by the dense atmosphere of the tube. That the arc or axis appeared along with the sky bodies before its radiance expired is to be determined in the next chapter, where its composition and operation are discussed.

CHAPTER SIX

THE ELECTRICAL AXIS AND ITS GASEOUS RADIATION

The binary electrical system was distinguished by an electrical flow between the principals. This was a highly energetic discharge which generated chemical, and probably nuclear, transformations among the gaseous constituents of the plenum. It provided the heat and energy for life to emerge.

The gaseous plenum initially seems to have contained an excess of hydrogen. The combination of electrical arc with hydrogen-rich gases favored the quick production of organic molecules and of biological systems (Miller and Urey) within the plenum.

An electrical current flowing along the axis of the binary of the order of 10^{14} amperes (100 => *teraamperes*) would produce sufficient magnetic field intensity to constrain the plenum gases up to 64 000 kilometers from the axis (see ahead to Chapter Seven). Electrical breakdown can occur in dense gases where the electric field intensity is of the order of 25 kilovolts per centimeter (Schröder, p. 90). The longer the electrical column the greater the potential difference required between the principals in order for breakdown to occur. Once breakdown occurs the voltage drop at a place decreases from kilovolts per centimeter to tenths of a volt per centimeter.[35] Along the discharge column the voltage drop varies considerably. It is greatest near the electrodes and is very small at most points in the body of the discharge.[36]

---

[35] The voltage drop occurs about a microsecond (one-millionth of a second) after breakdown (Bruce, 1955).

[36] Francis discusses conditions in the positive column of a short discharge tube. The estimate of 0.1 V/cm given above is a simplistic linear extrapolation from the data given for the voltage drop across an entire discharge tube. Actual values in the discharge are difficult to measure (Juergens, 1977a). In

It is difficult to estimate the energy that would be released by a discharging arc unless the voltage drop across the arc is known. Since we do not know this value for Solaria Binaria, we must define the arc's parameters in terms of other criteria. One must be preoccupied with the thermal constraints upon the Earth and its developing biosphere.

It is known that transverse heat flow is hindered by the presence of a strong magnetic field (Kapitza, p962). Also the gases insulated, diffused, and rendered uniform the intermittent blasts of the arc by the time the radiation reached the region occupied by the planets. Nevertheless, the closer the Earth to the arc, the more energy it would have received. For some time the Earth gained its energy almost entirely from the arc source. It is reasonable to assume that Earth's temperature must soon have devolved below 325 K. If the Earth-to-arc distance is chosen to be fixed at 64 000 km from the arc, the Earth is irrotational with respect to the arc, and it absorbs all incoming energy (see ahead to Chapter Thirteen); the heat flowing away from the binary arc cannot much exceed 2.6×10^{14} watts per kilometer of arc. Such an arc would dissipate at least 3.1×10^{22} watts within the sac. Given the constraints to radial heat flow cited above, the actual arc could have been much more energetic than we calculate here.

Thus, the discharge should have produced a small region of hot gas centered along the electrical axis. Surrounding the gases of the discharge was a large opaque mantle of cool gases. Within the cooler gases were the electrically charged planets, which had been repelled from the arc but caught up in the magnetic tube (see ahead to Chapter Seven).

In terrestrial lightning the period of electrical build-up (leader process) compared to the time of discharge (return stroke) is in the ratio of hundreds to one. The recovery time before the next stroke has built up is often 800 times the duration of the stroke. Thus it would be reasonable to conceive of the Solaria Binaria arc as discharging about one-thousandth of the time.

A lightning-bolt leader moves about 300 kilometers per second. Thus late in the history of Solaria Binaria it would have taken about 350 000 seconds[37] for the leader to work its way along the 105 gigameters between the principals. The return discharge propagates faster, taking only 2190 seconds or 36 1/2 minutes.

Did this arc, so necessary to life, persist even into the time of human

---

the plasma away from the electrodes the voltage drop is minuscule and could be one thousand times less than the average value.

[37] 4.06 present Earth days (sidereal).

awareness? As Super Uranus receded from the Sun, and the planets redistributed themselves farther apart in its wake, it is logical to assume that the intensity of the arc declined and its flow became intermittent. Hence, at around thirteen thousand years before the present an observer on Earth would have seen a great flickering and coiling axis or column of fire.

Solaria's electrical binary connection differs from a terrestrial lightning stroke of today in that it involves many concurrent (but not necessarily simultaneously launched) arc channels. A close analogy would be the granular cells seen at the bottom of the discharge channels between the Galaxy and the surface of today's Sun. In this latter case the difference between the arc in Solaria Binaria and the radially directed discharges on today's Sun is the absence of a closely spaced non-electrically neutral companion body. This proximity, which was present in Solaria Binaria, induced a continual series of electrical explosions to be conducted along the electrical tube joining the closest localities of the surfaces of the two stars.

The plenum gases at this time, especially near the arc, were dense enough to be opaque to radiation. A discharge that is opaque appears to radiate from its surface rather than from the whole volume of gas. In consequence energy flows diffusely away from the central discharge into the surrounding gas, some as radiant energy, some as the flow of excited matter, some by thermal conduction (by kinetic energy exchange in collision). Collisions will act so as to maintain an outward flow of energy (Somerville, p42).

Usually ions and electrons diffuse radially from the column. Later they recombine giving up the energy of ionization to the gas. Also, excited atoms, especially those which are long-lived, flow away from the column carrying internal excitation energy which they can release when deactivated by a collision with a non-excited atom or molecule.

The relative importance of radiation when contrasted with conduction for redistributing the arc's energy will depend upon the composition of the gas and the gas pressure. Some gases, like hydrogen and helium, are not efficient radiators of visible light. However, for all gases, high pressures make radiation more important than conduction in the transfer of energy.

In the laboratory, electric-arc current flows are of the order of 10 amperes per square centimeter. If such an arc were to flow between the early Sun and its close companion, Super Uranus, it need strike only a rather small area of the latter. A discharge column, if encompassing an area of only 10^{13} square centimeters, would produce an arc current of 100 teraamperes.

In an electric arc the gases "burn" [38] in a relatively narrow column. The higher the gas pressure the narrower the discharge column and the more difficult it becomes to sustain a uniform current through the discharge (Somerville, p. 19). Strong arc discharges, such as lightning channels, seem to bend into a helical shape. Such bending seems to generate a condition within the arc which can terminate the discharge (Blevin, 1964b, p473, Somerville, p54). "Non-electrical" gradients in the conducting electrified gases are usually offered as explanation for the curving of the discharge channel. These "mechanical" drifts set up within electrical discharges are probably better explained as electrical drifts, but neither explanation goes very far at present.

Revolution of the gases around the longitudinal axis of laboratory discharge columns tends to stabilize the discharges.[39] When they do, the rotating gases are said to create a radial "gravitational" field (Somerville, p. 20). Similar vortical stabilization is noted in rotating air; it is suggested in tornadoes where almost continuous vortical lightning activity occurs (Chalmers, p. 340).

The rotating gases surrounding and driven by the magnetic tube in Solaria Binaria would act to keep the electric discharge going when it otherwise would have gone out. In the laboratory, high current discharges are so unstable that continuous operation is not easily maintained. The pinch effect usually extinguishes the discharge. With the current removed, magnetic field relaxation occurs, so that the hot electrified gases begin to diffuse away, cooling the discharge column. Electrical forces quickly re-establish the current, stopping the outward flow of hot matter. So it was too in Solaria Binaria; the arc pulsed regularly responding to some natural rhythm between the forces leading to extinction and the forces promoting resurrection.

High gas densities favor brief, frequently recurring, pulses of arcs (Somerville p. 55). Could this mechanism be the origin of the regular pulses of radiation observed in celestial objects called => *pulsars?* As the gas density decreases, the arc's pulsing frequency would decline; pulsars show a slowing of the pulse rate with time (Hewish, p. 1083).[40]

---

[38] The gases in an electric arc do not burn in the sense of combustion, rather they are excited electrically, sometimes giving off light.
[39] The Gerdien arc, where stationary gas is surrounded by a rotating flow of water, shows very marked peripheral cooling, enabling a high axial temperature to be attained.
[40] Besides pulsing at intervals of one second or less, pulsars also show saltatory changes, named glitches

The electric arc operating in Solaria Binaria is a cosmic discharge of long duration. Bruce, in the course of seventeen letters about Cosmic Electrical Discharges (1958-1964), has documented examples of smaller arcs of shorter duration and of longer arcs lasting to millions of years. We concur in his conclusion that electric discharges on a cosmic scale explain many phenomena observed in the astronomical realm. Bruce has convinced us that only scale differentiates lightning discharges, observed regularly in the Earth's troposphere, from solar flares, periodic discharges in the giant envelopes of gases surrounding certain variable stars, and the enormous eruptions moving through the entire volumes of certain "active" galaxies (see, 1966a).

He proposed that an electrical discharge liberating energy comparable to that ascribed to the => *quasars* was capable of transforming elliptical galaxies into spirals. It would seem that the quasar phenomenon is in fact a galaxy in transformation. This is the grandest of the cosmic lightning discharges; in its wake the spiral arms of the galaxy form with their "metal-rich" stars. Bruce speculated that in the enormous temperatures generated in these discharges, nucleosynthesis transmutes smaller atoms into larger ones. It is this latter possibility that leads us to postulate that nuclear transformations were accomplished in the arc of Solaria Binaria. If they were, they probably occurred most vigorously at the beginning, when the discharge current was greatest.

Despite the many problems with laboratory experimentation in this area, some supportive work has taken place. Using a pulsed high current arc discharge, Russian workers produced beams of 40 kiloelectron volt => *deuterons* at instabilities in the discharge (Somerville, pp. 55ff). This achievement is consonant with certain proposed nucleosynthetic processes that occur in low energy flares above star surfaces (Canal). Zirin gives a mechanism for the generation of solar flares resembling processes which might occur within regions of a pinched electrical arc.

Even more closely related to the situation in Solaria Binaria is Joss' speculation that X-ray burst sources result from thermonuclear flashes. X-ray burst sources are episodic; in some, bursts are much more frequent. Many burst sources can be inactive for weeks. X-ray sources, steady and bursters, are associated with binary star systems. If, particularly, the burst sources are due to thermonuclear reactions in close binary star systems, then we can be confident that these reactions occurred in Solaria Binaria

---

(sudden decelerations of the object astronomers presume to be rotating). In the event that the pulsations are discharge phenomena, as we presume here, the => *saltations* could result if sudden outbursts altered the gas density irreversibly within the discharge column.

and that they were instrumental in shaping its chemical and biological structure.[41]

Inasmuch as these thermonuclear events were part of the earlier history of the electrical axis, no human would have observed this part of his ultimate creation. The last chapter mentioned what the earliest true humans would have generally perceived, but it postponed treating their special experience with the electrical axis. A correlation of the electrical axis with early legends about a central fire may be probative.

At one region of the Earth, the axis might be expected to appear as a kind of rainbow of fire or "neon-tube" glow across the sky ending at Super Uranus. In another region the arc might appear more short at the horizon and stretch to the red star. In the opposing hemisphere, the arc might be visible alone, first, and then might reach for and finally attain the Sun, with the axis blossoming at the Sun, thus "creating" it, or vice versa. The flickering of the arc, when slowed down enough to be noticeable, might resemble red coiled snakes, intertwining and crawling brokenly towards the great red god.

The snake and dragon accompany very early gods and goddesses. "The Serpent of the Jupiter-type myth is always seen to be a creation of the proto-Saturn god" (Tresman and O'Gheoghan, p. 39), that is, Uranus. The Saturnian image with snakes from India and the Chinese painting of the espoused deities, shown in Figure 12 and Figure 32, are suggestive. Serpents are among the earliest symbols of art and myth. The color red is widely used and sacred in archaic, perhaps Paleolithic Uranian times (Wreschner). However, the abundance of such symbols is countervalenced by their generality as referents. Lacking specific applications to phenomena, they are unreliable indication of the electrical axis. Certain symbols associated directly with Saturn (of the time of => *Super Saturn*) are also suggestive of the arc. These include the courtly long-gowned figure of the god, the tree of life (including the Christmas tree), the sacred mountain, and others (Talbott, D. N., ch. 8) that convey the image of the god atop a cone-shaped or pyramidal design on top of the World.

In Iroquois legend, at the beginning of things, the Chief of Heaven, in

---

[41] See behind, Chapter Two, where we argue that thermo-nuclear fusion does not occur in the interior of the stars. Theoretical models for the interior of solar type stars lead to the conclusion that their interiors, even if compacted, would not be hot enough to initiate nuclear fusion (Milton, 1979). Notwithstanding any contradictory calculation, the paucity of neutrinos emitted by the Sun must be considered as fatal to internal nucleosynthesis in stars (Juergens, 1979a). In solar flares and the other discharges mentioned below, temperatures significantly higher are measured. Thus only in the cosmic discharges does nucleosynthesis occur.

a fit of jealousy towards his spouse, uproots the tree whose flowers illuminate the celestial world. The Sun and Moon did not exist at the time. He cast his wife, "Fertile Earth", into the hole and replaced the tree (Eliade, 1967, pp.146ff).

Figure 12
The Planet Saturn in Ancient Indian Art

Brahma, the planet Saturn, encircled ring-like by serpents, testimony from an early time of the serpent motif in cosmogony. -- reproduced courtesy of S. I. S. *Review*

Among the Nagdju Dayak of Borneo, the Creator couple, dwelling as birds in the tree of life, fight and damage the tree badly. Some time after the first humans are born of their efforts, the tree is destroyed (*ibid*, pp. 77ff).

More closely correlative with the axis in the linguistic frame of modern science is the concept of the "Central Fire" that occupied early Greek philosophy. This has particularly descended through the fragments attributed to Philolaos, the Pythagorean (Dreyer, p. 40-3). Rose has thoroughly explored the material. Philolaos was the first of the secretive

Pythagoreans to publish a book and his treatment by Plato leaves little doubt that he represented a considerable school of archaic science.[42]

Some thirty-two attributes of the "Central Fire" are to be elicited from Philolaos and Heraclitus, as presented by Rose, all of which can be accommodated to the theory of the electrical axis of Solaria Binaria. The Central Fire was thought to have been a layer of fire above a layer of air. It is the center of the world. It is cone-shaped. It never sets, and has always the same location in the sky. It is "alone", the "highest", "unmoving", "stable", the beginning of everything. The Earth orbits around the Fire, but Earth does not rotate upon itself. Nor has the Earth any obliquity.

The Fire is not counted among the numbered bodies of the celestial sphere. It is not called Saturn or by any other name except that it was termed the "Mother of the Gods". It is called the "hearth of all", the "residence" of Zeus, his "throne", his "tower", his "fortress". It is a "divine ruler and teacher. It is the "altar", the "bond", and the "measure" of nature. The Sun borrowed light from the Fire; the Sun orbits around it.[43] The Moon, planets, and stars orbit the Fire.

Heraclitus reported it as "an ever-living fire, kindling itself by regular measures and going out by regular measures". He said that "it advances and retires" (Rose, 1979, p. 26).[44] Earth turns always the same face towards the Fire. A Counter-Earth exists, which is closer to the Earth than to the Fire, and obscures the Fire from view.[45]

The match of Solaria Binaria's axis of "electrical fire" (as electric discharges were called until the nineteenth century) with the attributes of the Central Fire in Greek cosmogony is close. The mention of celestial bodies can be explained as reflecting later observation of some traits.

---

[42] Rose supposes that the Central Fire is Saturn, the planet, as it anciently functioned, with which interpretation of the data we disagree, believing that the evidence is heavily in favor of its identification with the electrical axis.

[43] At a late time the Sun would appear to orbit the axis, as would Super Saturn, when these globes would appear to rotate around the point of the axis cone striking into them, as the Earth moved in its orbit around the axis.

[44] The Fire might advance and retire optically as it flared on and off in its decaying state.

[45] This probably is a phenomenon that followed the beginning of Earth's rotation perpendicular to the ecliptic, or refers to the era when the arc was no longer visible (see ahead to Chapter Fifteen).

CHAPTER SEVEN

THE MAGNETIC TUBE AND THE PLANETARY ORBITS

The arc along the axis between the principals created a magnetic tube, which surrounded the discharging gases (see Figure 13).

A magnetic field surrounded the electrical axis and extended outward to infinity.[46] The magnetic surfaces are here represented by lines, which by their increasing thinness indicate progressively weaker magnetic fields; see Figure 14. The strength of the magnetic field at a given distance from the axis depends only upon the magnitude of the electrical current flowing between the principals.

The ability of the magnetic field to constrain the motions within a gas depends upon the presence of electrified atoms. Whenever the energy density of the magnetic field at a given location exceeds the energy density of the gas,[47] the field can influence the flow of the gas and thus delineates the boundary of the magnetic tube. As noted earlier (Chapter 5) the presence of even a small fraction of an electrified gas can be sufficient to trap the neutral gases.

The electric current is mainly ions moving from the Sun to Super Uranus. It would be a negligible fraction of the total gas flowing between the two stars. Most of the electrical current was confined to small channels within the region of the flowing gas. Gas continually left the Sun and entered the plenum.

---

[46] A magnetic field only appears when relative motion exists within systems containing electric charges (Sherrerd).

[47] The energy density of a gas depends upon the density of the gas => *particles* and upon their temperatures (average random motion), and upon their kinetic energy as they flow within the electric field.

Figure 13
Magnetic Field Associated with an Electrical Flow

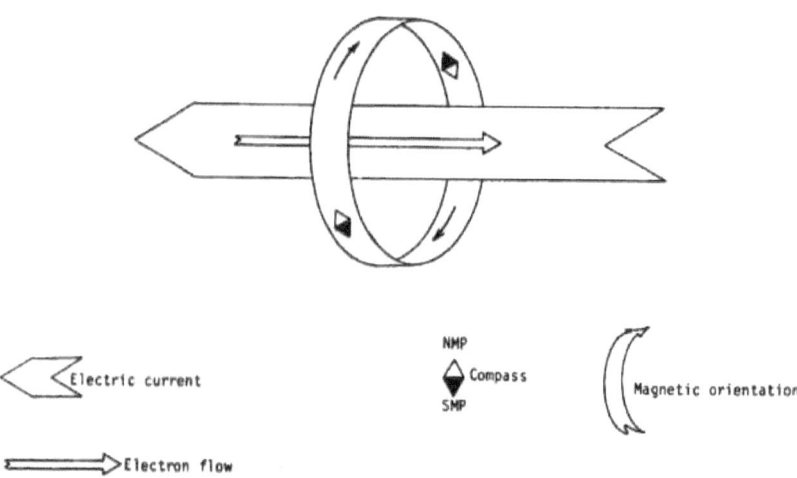

An electric current is always encircled by a magnetic field. By convention the direction of the electric current is opposite to the motion of the electrons contained in that current. For a moving electron the magnetic field is directed such that, if the electron flow follows the thumb of the *left hand*, the north magnetic pole of the magnetic field created by the electron flow is orientated around the motion in the direction of the curled left fingers.

Figure 14
Decreasing Magnetic Field Strengths
Surrounding Central Current at Increasing Distances

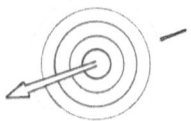

The magnetic field created by an electrical flow is oriented in the plane perpendicular to the direction of the electric current. The intensity of the magnetic field depends upon the magnitude of the current and inversely upon the distance from the current to the place where the intensity is being monitored. The further one is from a given current the weaker the detected magnetic intensity.

Most of this gas followed the electrical arc through the plenum (Alfvén, pp. 433-435). At Super Uranus the flow impinged upon a small area of the facing hemisphere. It is difficult to make direct observations of gas exchange within binary star systems, but some have been made. Batten (1973a, p. 2, p. 5) reports a typical value of 450 kilometers per second for the velocity of flowing gas. Using his value, we estimate that in the Age of Urania the flow may have amounted to about one-thousandth of the number of molecules in the plenum, or one hundred-millionth of the solar material per year.

In the region where the discharge passes, the gas would be hottest and hence of slightly lower density than that of the surrounding region. Moving outwards, the gas would become progressively cooler. If no other factors influenced the gas, we would expect to find gas density increasing in successively cooler layers. However, since the magnetic field grows weaker moving outwards, the density of gas that is constrained also drops. Thus, the highest gas density would be found in the warm region surrounding the discharge. Here marked chemical changes within the gases of the plenum are expected.

Because the electric discharge took the form of a pulsating arc, electrified gases could move radially during the relaxation cycle of the discharge. Gases of lighter mass move more rapidly than heavier gases and thus migrate more readily. Those atoms whose electrons could be most easily stripped off also migrated. This migration, coupled with chemical processes, altered the mixture in the magnetic tube until the gases now commonly found in the planetary atmospheres dominated it.

Given a 27-teraampere current flowing late in Solaria Binaria, the magnetic tube had the capacity to contain a gas density comparable to that of the Earth's present atmosphere (at surface level). The full plenum, at this same time, could have contained more than the equivalent of one hundred "Earth-masses" of gas and vapors.[48]

In a magnetic field, electrified atoms are constrained and follow the magnetic field direction at each location. Motions along the field are unimpeded, but motions across the field produce forces that cause an electrified particle to revolve round the local magnetic field line.[49] Combinations of along and across motions produce a spiral path about a magnetic field line, as shown in Figure 15.

---

[48] The "mass" of the plenum depends upon the gas composition.

[49] The authors understand that magnetic field lines are only a method of visualizing motion of charged particles being acted upon by magnetic fields and that they are not ingredients of the theories of electromagnetic interaction, but we feel that the use of field lines provides the reader with insight into the direction of the magnetic field around the electrical arc and into the motion that would occur as ions and electrons moved within the plenum.

Figure 15
Motion of Drifting Charged Particle in a Magnetic Field

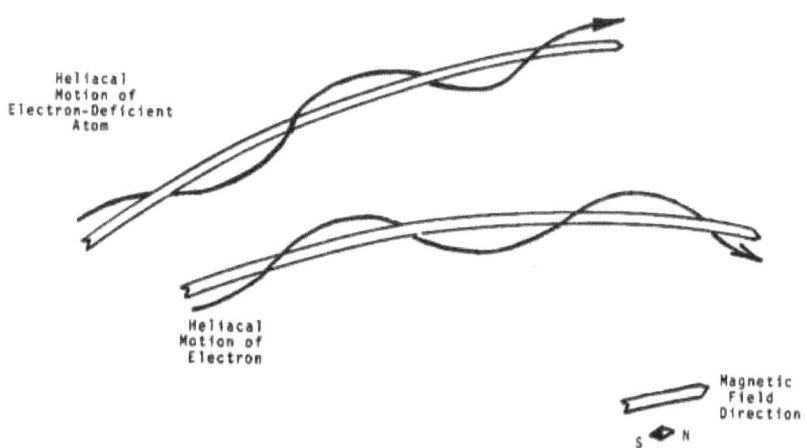

An electrically charged particle moving in a magnetic field is subjected to a force acting perpendicular to both the direction of its motion and the direction of the magnetic field. The result is that charged particles move freely along a magnetic field. The motion of the charge becomes helical because the constraint forces the particle to circle around the magnetic field while the particle translates along the field. Electrons and ions spiral in opposite directions.

The circular magnetic field surrounding the axial electrical discharge was also responsible for the production of much electromagnetic radiation within the plenum. Radiation would be emitted as ions and electrons were forced to spiral around the curved magnetic field lines of the magnetic column. This radiation process is called *Bremsstrahlung*, or braking radiation.[50] It is emitted whenever charged particles are retarded (decelerated). By radiating, the particles lose energy to the surrounding gas. An enormous glowing gas cloud soon surrounded the arc - an => *afterglow* - like the great ion trails left by large meteors.

As they lose energy, the spiraling particles move with smaller and smaller radii around the magnetic field line. By emitting braking radiation, motion across the magnetic field is greatly reduced. Since motion along the magnetic field is unaffected by the presence of the field, it eventually dominates. Thus the gases surrounding the discharge tended to flow around the magnetic column.

---

[50] *Bremsstrahlung* is observed from X-ray tubes, particle accelerators (synchrotrons), atomic beta decays, "supernova remnants" and cosmic X-ray sources.

The same radiation as described here is used by biologists to mutate rapidly growing species such as *Drosophila*, the common fruit fly. In Solaria Binaria, radiation was abundant precisely when needed to explain periods of great biological change.

Figure 16
Braking Radiation Emitted by a Spiraling Electron

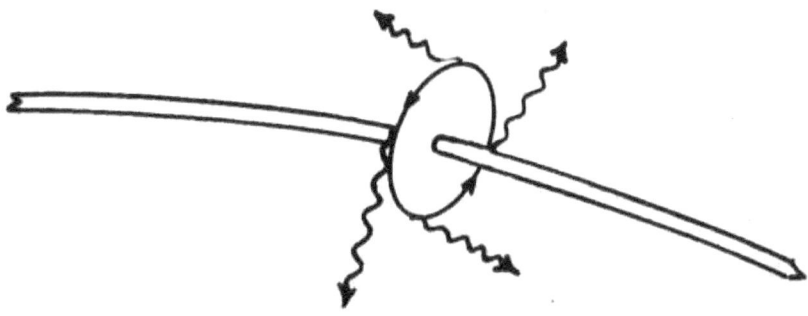

A charged particle experiencing a force accelerates; it gains or loses energy through the acceleration. When energy is liberated (because the accelerating particle is losing energy) it appears as electromagnetic waves emitted in a direction perpendicular to the charge's acceleration.

Earlier, we suggested that, in the outermost parts of the magnetic tube, material was revolving about the electric arc. This revolution began early in Solaria Binaria when the intercompanion current was greatest. The immense magnetism so generated was able to magnetize all of the contained and revolving material. Once magnetized, directed flow was assured (see Figure 13). Among this material were the electrically accreting primitive planets, also strongly magnetized. They revolved about the arc locked in direction by their magnetized structures.

As the binary extended, the arc weakened, in consequence of which the magnetism around the arc declined. With diminished magnetism, more and more material could exist in a non-magnetized state. Although this material had all been magnetized earlier, the magnetization would begin to decay as soon as the surroundings allowed it. The decay of magnetized Earth rocks has been documented by Nagata. There is probably a positive correlation between the ease of magnetization of a material and the duration of its remanence. So, as the magnetic tube weakened, the now orbiting bodies would lose their magnetism differentially depending upon their composition. As we will show later, Earth's decaying magnetism of today is a remnant of its stay in the magnetic tube.

Once in orbit about the arc the gases and solids (including the planets) of the plenum would retain their motions unless disturbed. This leads us to conclude that, in addition to their binary "dumb-bell" orbit, the Earth and the other planets also initially orbited in a circle about the Sun-Super Uranus arc (Figure 17).

Figure 17
Primitive Planets in Orbit About the Electric Arc

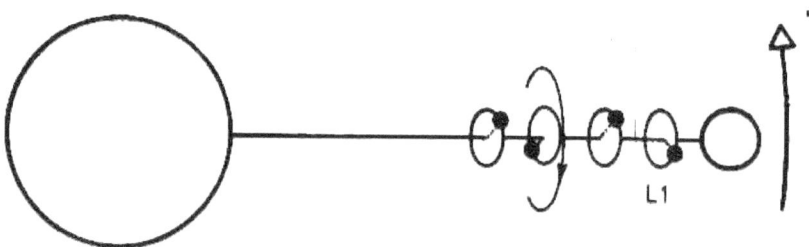

Over most of time the solar planets have orbited locked between the component stars of a binary. Their motion about the electrical connection between the stars resulted because the strong magnetic field generated by the electric arc kept the electrically charged planets in orbit around the arc. In a "gravitating" system only the Lagrangian point (labeled L1) allows a planet to co-revolve with the pair of stars, but for orbits under the influence of electrical force co-revolution is possible at many points along the electrified axis of the system.

Because of electric repulsion, we propose that the arcuate orbits of the primitive planets were situated somewhere in the vicinity of what is called the => *Lagrangian point* L1 for the Sun-Super Uranus binary system. Near this point co-rotation is expected as the dumb-bell rotates. In fact, even in today's Solar System, where the magnetic tube has collapsed, a Sun-Earth-satellite pair (ISEE 1 and 2) has been orbited at the L 1 point between the Sun and the Earth. In today's system, all the planets orbit with different times and the L 1 orbit is barely stable; but in Solaria Binaria, with a strong magnetic tube in place between the principals, L 1 would be a most likely haven for planetary bodies in the binary.

Since we propose the existence of more than one primitive planet, and since each is electrically charged, the planets would revolve about the arc staying as far from each other as the principals would allow. This, we feel, would crowd the four planets near the L 1 point.

The planetary orbits about the axis, like rings on a pole, would be substantially closer than today's concentric orbits. The planets would maximize their separation, subject to three constraints: repulsion by the principals, repulsion from the arc, and the need to follow the magnetic lines of the tube.

In consequence each planet orbited about the axis tending to expand or contract its orbit depending upon its charged state relative to the axis and its need to stay away from the other planets. Their speed and direction was dictated by the flow of magnetized material about the arc, despite their need to avoid one another. Given that the planets as a group occupied a limited region of the tube, they positioned themselves on their orbits so as to maintain the net maximum distance from the summated repulsion of all of the other orbiting planets. Perhaps the simplest, and at the same time adequate, response to the constraints would be for each planet to take a different azimuthal position[51] on its orbit in the magnetic tube.

In the opaque plenum, and because of their different revolutional phases, no planet could be observed from another planet. Even when the plenum was clearing, the planets so positioned would be difficult to discern. They may not have been perceived until the => *Age of Jovea*, when, freed from the tube, their dim radiance could be isolated against a darkened sky.

The magnetic tube may have played a part in generating cosmic sounds. Archaic Greek philosophers, especially the Pythagoreans, employed the phrase "music of the spheres" to designate what has since been regarded as an unreal belief in celestial and planetary sound. That the violent forces within the tube would have emitted acoustical waves is unquestionable.[52] In the manner of => *whistling atmospherics*, as lately studied (Hines, p. 816), such sound would be trapped by the magnetic field and propagated along the magnetic tube.

In the late times of the tube, when celestial bodies could be distinguished visually, the sound might have inspired the Pythagoreans to the invention of their sacred musical scale, which was also related to their sacred theory of numbers - both sound and numbers constituting theophanies. The sounds might first have been involved in the earliest sacred music. The magnetic tube worked its wonders by an invisible hand.

---

[51] Azimuthal angle is, here, a measure of the planet's progress around its orbit. Planets whose azimuthal angles differ would be said to have a revolutional phase shift (as in Figure 17).

[52] Several expert observers, working at remote locations, report sound associated with intensive displays of the Aurora Borealis (Harang, Stomer).

CHAPTER EIGHT

THE EARTH'S PHYSICAL AND MAGNETIC HISTORY

The generally round shape of the Earth is an effect of external electric pressure to bring it into electrical balance with the plenum. It was originally a dense aggregate, a fragment of Super Sun trapped in the magnetic field that was generated around the arc joining the Sun and Super Uranus. The aggregate grew rapidly in a short time from accretions of smaller bodies and chemical elements.

The Earth's density alters from lighter material on the outside to heavier on the inside, in proportion to the intensity of requirements of materials for charge - lesser on the outside, greater at the center. Density and conductivity are correlated. In an intensely electrical ambience, the deposition and transmutation of metals such as iron and nickel at the core of the Earth are understandable.

Though knowledge of the Earth's interior is by inference, its seeming simplicity may be a fact and, if so, the result of its electrical accretion and its conductive nature. By contrast, the superficial crust of the Earth consists of more poorly conducting species. It has also been subjected to many geophysical incidents of a recent kind. Therefore, its highly differentiated structure is understandable (see ahead to Chapter Eleven). Granite, for example, is a rock formed as a global covering at and near the surface under highly energetic conditions. It may once have been basalt that was electrically energized by the magnetic tube to the point of metamorphosis. Too, it may have migrated by electrophoresis and deposited by electrolysis, as particulate, from the enshrouding plenum. It is old, then, but not so old as the core and mantle of the Earth.

Granting that the granite cloak could not be a metamorphosis of sedimentary rock requires admitting that the sediments can never have been very deep, not much more than the observable sedimentary cover on the continents and ocean bottoms today! A half-million years of violent and gradual erosion would seem to be sufficient to provide it. If, as will be argued in Chapter Thirteen, the Earth has lost crustal material by explosion, it has also gained some materials from the explosions of foreign bodies (see Chapters Eleven and Fourteen); the search to explain ore, salt, and other anomalous bodies embedded in the surface must begin with a study of their possibly cataclysmic accretion.

The mineral structure of the Earth harbors magnetism, which is the capacity of some of the Earth's rocks and of its total surface and ionized atmospheric gases to give evidence of a distinctive electrical presence both now and in the past. Rock magnetism, imprinted in ferruginous and other rocks by some past event, yields magnetic intensities up to one microtesla.[53] At the surface the magnetic field of the Earth's body has an intensity of sixty microteslas. In rocket and orbiting satellite observations made in the ionosphere, high above the surface, intensities as low as ten nanoteslas are found. This ubiquitous force is weak to the point of impotency, yet at the same time it is highly significant in reconstructing the Earth's history and present state.

The globe as it accreted was aligned with the magnetic field lines around the electrical axis discharging between the Sun and Super Uranus. It was forthwith magnetized.[54] It, too, orbited the axis, maintaining a fixed direction relative to the magnetic field in which it moved. As depicted in Figure 18, it posted its rotational poles at right angles to its magnetic axis.[55]

Quantavolutions eventually weakened the field of the magnetic tube leaving today only a feeble magnetic field in the region of the ecliptic (see Figure 11). The outflowing solar wind protons seem to leave the Sun radially. Except near the Sun, this flow seems to be focused mainly onto a disc enveloping the ecliptic. The outer-planetary space probe Mariner 10 has noted some depletion of solar wind at high ecliptic latitudes (Kumar and Broadfoot).

---

[53] The unit of magnetic field intensity is the tesla. Such an intensity is very strong, comparable to the largest magnetic intensity noted in the cosmos. One tesla represents one hundred million magnetic lines of force passing through each square meter of the magnetized surface. The nanotesla is one-billionth (USA) as strong and represents the weakest detectable magnetic intensity.

[54] As were all planets then in the tube, meteorites are generally found to be magnetized (Levy). The cases of other bodies will be treated later.

[55] This rotation would have the same period as the Earth's revolutional motion about the electrical axis. The poles of rotation would lie parallel to the arc.

Figure 18
The Earth in the Magnetic Tube

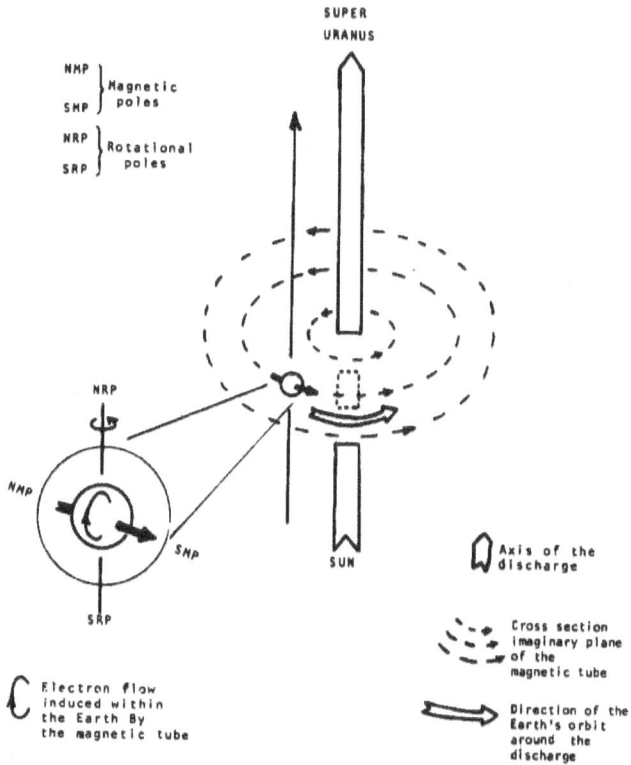

The magnetic field around the electric arc of Solaria Binaria made the electrically charged Earth orbit around the arc. The magnetic intensity of the constraining field caused the material of the Earth to become magnetized. So held in orbit the Earth's rotational and magnetic poles were located 90 degrees apart on the Earth's surface, the rotational axis was directed parallel to the arc, while the magnet axis was directed along the contours of the magnetic tube.

After sufficient weakening of the magnetic tube, the Earth was released from alignment with the field lines surrounding the electric arc; gyroscopic action, caused by the electric current flowing through the Earth's core, then ended the former rotation about poles displaced greatly from the magnetic axis. The Earth-magnet sought alignment with the now dominant solar magnetic field created by the motion of the electrically charged Super Uranus around the charged Sun. The Earth began to rotate with the north rotational pole (geographic north) in the same place as the "north" magnetic pole (Figure 19).[56] Later events separated the two poles.

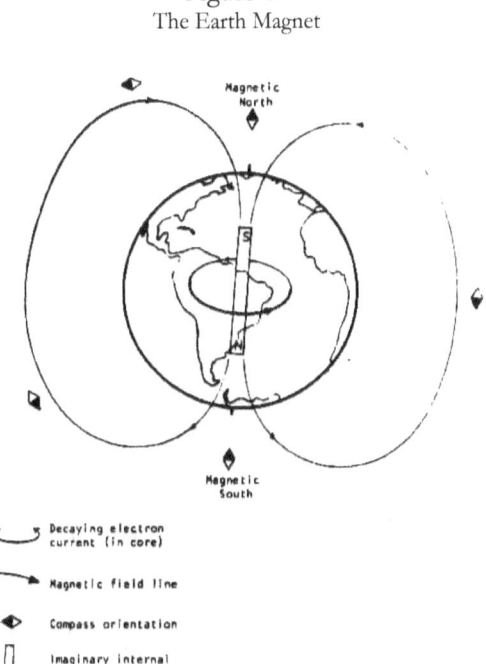

Figure 19
The Earth Magnet

The surviving magnetization of the Earth's interior arises from the remnant of the electric current induced within the Earth's material during its stay in the magnetic tube. This decaying current is detected externally by the presence of an ever-weakening magnetic field, which surrounds the Earth.

Presently the magnetic axis is tilted eleven degrees to the rotational axis (Haymes, p214). The term *far-magnetic field* refers to this dipolar field observed from a great distance above the Earth. The *near-magnetic field* has its poles in northern Canada and on the Antarctic coast, south of Australia. Here the magnetic field is vertical. Location of these, often called "dip poles", is difficult and somewhat dependent upon crustal conditions rather than upon the internal magnetization (Haymes, p. 217). The present dip pole in the northern hemisphere[57] drifts westward by more than five

---

[56] This is the south pole of the internal magnet. It attracts the north pole of a compass.
[57] The north dip pole is located between Bathhurst Island and Prince of Wales Island in the Canadian Arctic (260° E, 74° N). Its motion is complex but reasonably well documented since 1950 (Dawson

kilometers per year (Vestine, p.90). Its daily motion carries it through an elliptical loop with amplitudes up to 130 km reported (Serson).

Only in the recent quantavolutionary periods (the post-Saturnian: see Chapter Fourteen) have the magnetic poles abandoned the equatorial region. Palaeomagnetic estimates of the location of the ancient magnetic poles of the Earth's surface register an aversion to high latitudes (Lapointe et al.).

Under earlier Solaria Binaria conditions, therefore, the surface rocks and internal magnetism of the Earth were in line with the field forces of the magnetic tube. All subsequent accidents to the Earth that brought magnetic disturbances, whether in the rocks or in the poles, must be overlaid on the fundamental magnetic map imprinted upon the globe during its youth. Furthermore, the electric generator of the Earth's magnetic field must be the descendant, still declining, of the primeval current set in motion by the magnetic tube. This current flows in the conductive material deep within the Earth. There it creates, and mainly defines, the field, the lines, and the poles of today. Its ancestor, much stronger, was present to imprint magnetizable rocks under circumstances of changes.[58] Today, many rocks point magnetically towards what was some pole of the past, some to the neighborhood of the present magnetic pole, and most to nowhere in particular. Only a few of the rocks are magnetized at all.

The magnetic poles of today are located near Thule, Greenland, and in Antarctica (120°E, 75°S). When these poles are joined, it must be noted that their axis does not transect the center of the Earth - it is offset by 436 kilometers towards the surface of the sphere, where lies the basin of the Pacific Ocean (Haymes, p. 214). From this it may be inferred that, subsequent to the establishment of the magnetic field of the Earth, a quantavolution scooped out the Pacific basin and deformed the Earth (see Chapter Thirteen).

The present global field, which we have said is descendant from the Earth's stay in the magnetic tube, is complex in that later events have acted either to induce new electric currents (located superficially within the core) or to perturb parts of the main current flow. The result are the disturbing currents, shown in Figure 20, the imprint of more recent quantavolutions of the world order when Earth suffered electrical encounters on a large

---

and Dalgetty) with some data over the past millennium (Yukutake).

[58] It is known that molten rock will be imprinted if it solidifies, and then cools to its => *Curie temperature* in the presence of a magnetic field.

scale (de Grazia, 1981, 1983a; Juergens, 1974, 1974/5; Velikovsky, 1950, pp. 85ff), including meteoroid impacts (Dachille, 1978) and encounters.

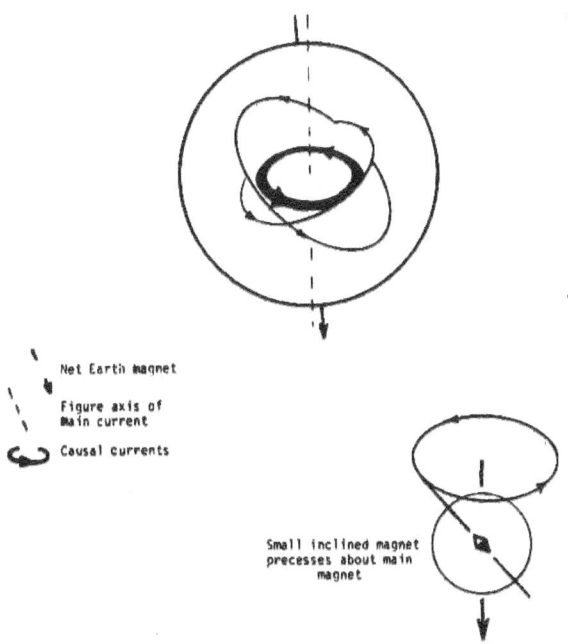

Figure 20
Magnetic Transactions Within the Earth.

The drift of the Earth's magnetic dip poles across the continental surfaces indicates the complex nature of the causal current through the material making up the Earth's bulk. It is likely that the major current (drawn thickly) was induced during the Earth's stay in the magnetic tube. However, lesser currents (drawn thinly and located closer to the planet's surface) were individually induced in each of the interplanetary encounters of the Late Quantavolutionary period. Each of these lesser currents must transact with the main current, and likely also with one another, resulting in a complicated precession of the total magnetic field around the figure axis, which is directed perpendicularly to the plane of the major internal current (see inset). The net motion is the observed drift of the magnetic dip poles.

Not only can surface anomalies be explained by celestial intrusion, but so can the wander of the dip poles, a vector sum of the complex wobbling. If large electrically charged bodies passed close to the Earth's surface they could especially disturb the electric current in the core as noted above. These lesser currents, once created, would interact with the existing magnetic domain of the Earth (see insert, Figure 20).

Malkus concludes that precession of the Earth's rotational axis produces torques upon the Earth's fluid interior. He sees these torques as generating the internal dynamo that is conventionally called up to create the Earth's magnetism. Here we adduce his results only in evidence of magnetic wobble arising from torques.

The Earth's magnetic field has been weakening over the 150 years of measurement of it strength (Cox, p237). This implies a decay of the current within the Earth's core. Such a decay could be the main source of heat flowing from the Earth's interior. At the observed rate of magnetic decline, it would take on the order of six hundred years to heat the core one (degree) kelvin. Even granting a much stronger field ten millennia ago we do not believe that the Earth's core is fluid. The observed surface magnetism and seismic profiles of the Earth's interior are consonant with a solid conductive body containing an excess of free electrons. Given that the Earth's field is weakening, it is logical to believe that rock magnetism is decaying at least as rapidly (see behind, Chapter Seven). Neither would still be present if magnetization had not occurred very recently.

The magnetic testimony of the lithosphere is largely fossil, in that the present interior current of the Earth passes its magnetic force into the atmosphere without the capacity for imprinting anything except molten rock. That is, if some rocks carry a complex magnetism, it must be measured and read as a much more intricate registry than the present magnetic field could generate.

As indicated earlier, the strength of the Earth's magnetic field is over fifty times that of the strongest rock magnetism. Presumptively, in the magnetic tube the Earth's overall magnetization would have been only a fraction of that of its environment. Notwithstanding its genesis the time measure of the current within the Earth's core is to be adjudged by the surface magnetic field and not by the rocks. Rocks containing => *magnetite*, of igneous origin, are imprinted by the Earth's field when they freeze. Other rocks containing similar minerals can be made magnetic if subjected by lightning to piezostress (Hertzler and Phillips or to magnetic shock (Dachille, 1978). Magnetization by any, or all, of these modes can occur when large charged cosmic bodies encounter the Earth.

Magnetic surveys disclose magnetic axes in all directions (Mil-som, Vestine, p94). Typically, the survey instruments are set to read as "north" and "the reversed north". That is, the preconceived theory calls for a magnetization in the direction of the (wandering) north magnetic pole, and, in recent years, evidence that the poles may be on occasion reversed, "north" thereupon reading "south". The theory is vitiated by lack of consistency in the readings. To revive the theory, extinct poles in off-north directions are postulated as the determinants of deviant readings, even though this practice begs the question by using two variables to prove each other. Juergens (1978) has criticized the interpretation of published evidence of geomagnetic orientations and reversals (see also Cox, p. 244).

The Earth's magnetic field has never been reversed. It is securely implanted in the Earth. Should the earth have tilted or turned upside down (Warlow), our model requires that its magnetic field would have turned with it, acquiring perhaps some minor dislocation or a tangential minor current as an offshoot.

Once the magnetization has stopped, the magnet decays. What is the duration of the Earth's magnetic field and its rock magnetism? Until recently both were considered permanent or assigned exceedingly long durations. Now it is recognized that magnetized objects lose their magnetism over intervals that are impressively short, Cook (1966, p. 282), using data given by Nagata, estimates the total decay time at under 70 millennia. By our theory, the magnetic tube would have held sway over the Earth's magnetic field and any lithospheric imprinting up to its weakening and collapse some 6000 years ago. If the tube were weakening, the Earth's field should have decayed with it. After the tube collapsed, the Earth's magnetism began to function independently. Its continued loss in strength has been noticed.

Barnes summarizes measurements made of the Earth's magnetic moment and magnetic field intensity from the determination by Gauss in 1835 until the middle of the decade past. These data show that the magnetic moment is decaying with a half-life of about 1400 years.

He notes that the energy in the Earth's magnetic field can produce, by self-induction, an electric current in the conductive core of the Earth. This current loses energy to the core in the form of heat, producing the observed decay of the external magnetic field. At present, by his computations, the core current required is 6.16 gigaamperes with a power loss of 813 => *megawatts*. If the Earth's field had been decaying undisturbed for more than a few thousand years, magnetization would have been present whose decay should have melted the Earth.[59] Recent onset of the presently noted decay seems in order.

From the Earth's magnetic moment and using Barnes' estimate of the present internal current, we arrive at a "radius" for the Earth magnet of

---

[59] On similar grounds, cosmogonists have rejected the possibility that the Earth's core contains its share of the radioactive elements posited as the Earth's cosmic allotment.

two megameters (about one-third of the globe's size). Since the magnetic intensity at the surface is a dilution of the internal magnet, discussion should be focussed on the latter. Our estimates yield a magnetic intensity close to ten times the surface value at the source. The decay of this magnet over the past few millennia is of interest, for, adapting the decay calculated by Barnes, we obtain the data in Table 4.

If no quantavolutions had occurred, the above extrapolations would predict that seven millennia ago, the Earth's magnetization was thirty-two times it present strength. In the same era, then, the heating of the core should have been 32 squared, or 1 024 times the 1970 value. Under this enhanced decay, the core would be heated by one degree in 226 days. This heavy heating could warm the iron in the core above its Curie temperature in five centuries were it to continue undiminished. Since several celestially-induced saltations punctuate this interval, it is unlikely that the magnetic decay can be extrapolated meaningfully back through the interval. Even if it could, the Earth core would still remain safely cool since the liberated heat is not all retained in the core; it flows outward towards the surface; and on its way it encounters over thirty times the volume of material of its region of genesis. The surrounding mantle material requires up to twice the energy per kilogram to heat as it does the metal-rich core. Thus the heat is easily dissipated providing the Earth-magnet is not allowed to grow further into the past and, indeed, this it need not do, for during its stay in the magnetic tube the current did not decay and its energy output was benignly dissipated.

Electricity probably played an important role in cooling the Earth's interior in the days of great magnetization. Evidence abounds that, under electrified conditions, heat flow and heat dissipation patterns are altered over those noted in the absence of electrical flow (see Asakawa). Earth currents persist to this day; we have no reason to believe that they were less strong in the past. Their role in shaping and maintaining a habitable globe cannot be overemphasized. We do not know the maximum magnetization during Earth's stay in the tube, nor its level when the tube collapsed, releasing the field to free decay. The level of magnetism induced in a magnetizable material depends upon the purity of the material, the temperature, and the strength of the inducing field. The Earth's core is unlikely to be a pure magnetic alloy, hence its magnetization in the tube would not have to reflect more than a small fraction of the full strength of the inducing field. On leaving the tube the core need not have been

Table 4
CALCULATED UNDISTURBED DECAY OF THE EARTH'S MAGNETIZATION

(using Barnes' Decay Model)

	Date (Astronomical Years) at surface	Magnetic Field Intensity (in => milliteslas) within core
+ 1970	0.062	0.61
+ 570	0.124	1.22
- 830	0.248	2.44
- 2230	0.496	4.87
- 3630	0.992	9.74
- 5030	1.98	19.5
- 6430	3.97	39.0

Reckoning in astronomical years. AD years are designated with a +, BC years are lessened numerically by one year and have a - preceding them. (e. g.: 1 BC = 0.; 3 BC = -2)

magnetized to any level that would pose a problem in thermal dissipation, whatever the model employed for the heat flow that began as the magnet waned.[60]

---

[60] The Earth's rotational spin-loss, ascribed to tidal friction, liberates forty-two million times the energy

Given a half-life for magnetic decay of the order of 1 400 years, it is reasonable to conclude that all existing magnetization of surface rocks must be very recent. A rock magnetized to one microtesla (about the strongest value noted) would decay to the limit of detectability (one nanotesla) in ten half-lives. If rock magnetism decays at least as rapidly as does the Earth's field, fourteen thousand years would erase all magnetic imprints from the rocks! Not only must the rock magnetism be very recent, but also most of it has probably resulted from electro-thermal events of cosmic origin.

The presence of magnetism throughout the Earth's domain cannot be denied, despite difficulties in explaining its generation and variation when using models which maintain that the Earth is not an electrically charged body. Those who have studied the electrical currents associated with the body of the Earth and the higher atmosphere above the Earth, and those who have studied the electrical flow from the atmosphere to the ground and its variation, might well have concluded that the Earth is most easily understood as an electrically charged body. That they have not so concluded is significant. From the earliest modern experiments in electricity the evidence of an electric Earth has loomed closely under the printed pages of explanations. Many investigators perceived the answer but were discouraged by their inability to offer proof of their suspicions (for example, Sanford, p. 105, pp. 72ff). Our assertion that the Earth is a body that carries a net surplus of electrons is paramount in understanding its properties.[61]

In the beginning the Earth was far from electrical equilibrium with the plenum of the young Solaria Binaria. Consequently the accumulating Earth material transacted strongly with its surroundings. The Earth probably glowed visibly as it formed and for a time thereafter.

At an early date this visible Earth-glow was extinguished and the Earth became the dark planetary body that it is today. An electrical current of 1800 amperes still flows from space to the Earth. This continuing electrical transaction partially decreases the Earth's charge by 3.5×10^{29} electrons per year. This altered charge represents a flux that is ten times that ascribed to the Earth-magnet in the core. The Earth-air current density is 3.5 microamperes per square kilometer of surface. There is evidence of a

---

presently lost by the magnetic field. The Earth has not boiled from the tides (compare with Darwin, 1879).

[61] We remind the reader that this electron surplus is relative to the Earth's material itself: relative to the cosmos the Earth is an electron-deficient body, while relative to its immediate surroundings the Earth is close to, but not quite at electrical equilibrium, as we shall note below.

possible electric connection between the Earth and the Sun; this circuit drives, in part, the Earth's weather cycles (Webb, Cole).

The energy liberated by the Earth-current is in addition to that from the influx of sunlight. Its power has yet to be determined and its significance is mainly unexplored. Nevertheless several phenomena are recorded indicating the Earth's electrical state. An electrical gradient exists, increasing the electrical potential maximally near the ground by a few hundred volts per meter of upward displacement (Chalmers). Higher, the gradient declines, producing a maximum potential difference of 300 000 volts between the ground and the atmosphere at an altitude of twenty kilometers. The direction of this gradient is consistent with the notion of a negatively charged Earth in a slightly less negative environment.[62] So the => *troposphere* forms an electrical sheath joining the ionospheric plasma to the charged Earth.

Above, in the => *ionosphere*, strong electrical flows are documented with maximum currents of the order of 90 000 amperes. These flows occur in a plasmasphere analogous in form but not in behavior to the Sun's photosphere.

Farther up, another electrical sheath, a => *double layer*, exists which joins the plasmasphere below to the solar wind above. This sheath, at the so-called *magnetopause*, has produced phenomena that have defied explanation (Kelley) because electric neutrality is demanded of the Earth. The double layered sheath, like the chromosphere-corona of the Sun, is the gatekeeper for the systems. It admits and accelerates incoming electrons, while it repels or retards incoming ions. From the Earth-side it prevents electrons from escaping and facilitates the outflow of ions.

On occasion, solar outbursts flood the double layer, diminishing its effectiveness (Hartline) and suddenly altering for a time the Earth's charge level. This produces a saltation in the length of the day, that elsewhere has been called a "glitch" (Danjon; Challinor; Gribbin and Plagemann, 1973). In the weeks that follow, the Earth regains its charge balance and the rotation corrects itself. Rotational saltations are explainable in terms of a charge exchange between the Earth and the surrounding interplanetary plasma.

Inasmuch as in the past the Earth was farther from equilibrium with its surroundings than it is now, electrical readjustment was more spectacular than the small electrical transaction noted today. As the Earth came into

---

[62] Such an arrangement of charges is seen elsewhere; it may be a means of shielding the Earth's electron complement from a voracious Sun (see Technical Note B).

balance it would appear to an Earth-bound observer that the Earth's electrical charge was decreasing with time, whereas in fact the opposite is more correct. The Earth is gaining charge continuously. In line with the electrical explanation for rotational saltations, the deceleration of the Earth's rotation is explicable as a charge increase with time.

We maintain that the Earth's very geophysical integrity is determined by its continuous charging and the interruptions thereof. There are links between volcanism and climatic change, and tidal phenomena are linked with both of the former and with seismicity (Roosen *et al.*). It is suspected that an extraterrestrial trigger is responsible for these correspondences (Rampino *et al.*, p828, Johnston and Mauk, pp266-7). That trigger is intimately related to variable rates of charge accumulation by the Earth. These variations have been in the past responsible for drastic quantavolution of the Earth's surface.

There is mounting evidence that even the biosphere is shaped in consonance with the Earth's electric and magnetic state. Discussion of this subject need not be further postponed.

CHAPTER NINE

RADIANT GENESIS

The physical history of Solaria Binaria may be divided into three major periods according to the intensity of quantavolution occurring: a primary period of violent changes and rapid development, extending perhaps to a quarter of a million years; a secondary period of relative balance among the elements within the system, extending almost to the present; and a shorter tertiary period of system breakdown, when Super Uranus, the planets, the sac and plenum, and the electrical arc with its magnetic tube underwent abrupt transformations.

A biosphere was generated during the primary period and produced its main forms. That is, there was first a time of radiant genesis, a proto-zoic stage, followed by a time of the escalation of basic biological types, a palaeo-zoic stage. Then occurs a meso-zoic period of formal and ambient stability, which coincides with the secondary period of relative balance in physical history. These are the subjects of the present chapter. The Cenozoic, which we redefine as a period of explosive quantavolution, corresponding to the period of system breakdown, is the subject of Chapter Twelve; there the origins of human nature will be discussed (see also Table 6).

The prevailing theory among scientists conjectures that a sequence of chance chemical combinations occurring over time produces the "self-replicating molecule" deoxyribonucleic acid (DNA). For the moment we pursue this idea of chance chemical combinations.

In Solaria Binaria, the sac is the vat of chemical evolution. Its gases are hydrogen-rich but contain, by inheritance from the body of Super Sun, all simple ingredients found in life forms. The energy sources which catalyze the process are ultraviolet radiation, electric discharges (lightning bolts), and ionizing particles (from cosmic rays or radioactivity). Using a variety of gaseous mixtures, energy sources and temperatures, experimenters have been successful in producing a multitude of prebiotic compounds in short times.[63] The ultimate step, the creation of life, has not been reproduced in the laboratory ! Presently, experimenters are searching vigorously for some means of reproducing the "reproducer" - DNA - in the laboratory.

The composition of the plenum gases varied significantly over time, though for a long time the gas density remained fairly constant. Once Solaria Binaria came into existence, electrical forces produced => *electrophoresis* among the electrified atoms throughout the system; in electrified gas mixtures the components apportion themselves within the mixture in relation to their ionization potentials. "The component with the lowest ionization potential becomes more concentrated at the => *cathode*, that with highest ionization potential at the => *anode* " (Francis, pp195ff). The rate at which separation of the constituents occurs depends upon the => *mobility* of the ions. The mobility of an ion is of the order of one to ten centimeters per second for each volt per centimeter of electrical field (at standard atmospheric temperature and pressure - S. T. P.). At constant temperature the product of ion mobility and pressure is approximately constant (Papoular, p. 94).

The least => *massive ions* are the most mobile and so they will migrate soonest; the heavier ions will take longer to separate. In Solaria Binaria only a partial separation was effected, but this was sufficient to contribute to the anomalously low abundance of lithium, beryllium and boron noted in the solar spectrum (Ross and Aller).

The effect of the discharge was to reapportion the plenum gas mixture, changing the local percentage of hydrogen relative to the heavier atoms. This would effect greater efficiency in producing organic compounds in certain regions within the plenum (Dayhoff *et al.*, p. 1462).

After the nova (see behind to Chapter Four) the plenum occupied a large volume; it was honeycombed with variously electrified domains producing a state of great electrical dis-equilibrium. Held together by

---

[63] The work of Stanley L. Miller and Cyril Ponnamperuma stands out.

pervasive cosmic electrical pressure, the gases of the plenum assumed the smallest volume consistent with their charge density. In reaction to the nova, electric flow within the plenum worked to equalize charge densities within the sac, while maintaining an outward radial gradient of increasing charge density in concession to the external demand from the continuing cosmic transaction.

The result was an initial implosion of the sac, as charges were redistributed, superposed upon a much slower expansion of both the sac and the rest of the system as galactic charge accumulated. Consequently, over most of their history, the Earth and the other primitive planets were immersed in a dense plenum of gases which was opaque to radiation; this gas was at least as dense as the present atmosphere at the Earth's surface.

The nutritive soup from which living forms emerged was not wholly the primitive vapors of Earth (conventionally the oceans and atmosphere) but the total surface of the planets and the volume of the sac. Appropriate temperatures were available in most of this volume within thousands of years of the nova of Super Sun. Various organisms can survive temperatures well above the Earth's present temperature. Fish, fly larvae, and aquatic metazoans survive in hot springs where temperatures approach 320 K (Dicke, 1964, pp. 119ff; Wickstrom and Castenholz). Live bacteria have been discovered in an oil well where temperatures approached the boiling point of water (Dicke, 1964). Thus it is argued that the Earth could have had a much warmer climate in ages past when life arose. Urey concludes that temperatures have been below 425 K since the Earth's crust separated (Miller and Urey). Fox (1960, p. 203, p. 206, 1970), maintains that certain chemical processes preceding the genesis of life were accomplished by heat. He now considers the debate over past temperatures irrelevant since the critical processes can occur at temperatures well below 425 K.

If we consider only that portion of the plenum which enveloped the planetary region (a cylinder 35 gigameters long by 100 megameters diameter) we have a reactor volume which is sixty million times the combined volume of the Earth's atmosphere and oceans, in which life otherwise is believed to have been generated. The energy source for the plenum was the electric arc. The early arc may have liberated about 10^{23} watts to the plenum, compared with 3×10^{13} watts received as ultraviolet radiation by the Earth's atmosphere,[64] or with 3×10^{11} watts received as

---

[64] Miller and Urey cite this value for radiation capable of modifying the primitive gas. The more complex molecules produced after the initial photolysis are more easily excited and are affected by

lightning discharges (see Chalmers for data).

If Solaria's plenum at the edge of the central flow zone is compared with the outer surface of the Earth's atmosphere with regard to energy density, Solaria's plenum will have had an advantage by a factor of 500 000.

At the other extreme, if the energy is spread throughout the entire volume of both reactors, the advantage in energy density still is with Solaria fifty-fold.

If the time taken to generate life in an energized primitive environment depends primarily upon the rate at which the primitive gases can be excited to produce chemical changes, then life ought to have been generated within the plenum after a time somewhere between two thousand and two hundred million years! [65]

Should the initial photolysis not be the rate-controlling step, then the immense volume factor greatly favors a more rapid biosynthesis in the plenum than supposedly occurred in the Earth's atmosphere and oceans aeons ago. Furthermore, a highly electric environment may speed up generation time, and therefore the intergenerational opportunities for mutation. As we see it, the plenum was an ideal reactor in which living systems could be synthesized and sustained.[66]

Evidence that the generative environment was highly magnetic can be inferred from the sensitivity of many living organisms to magnetism. Both animal and plant life respond to strong magnetic fields (above 100 milliteslas), showing modified growth or behavior (Kolin, pp. 40ff). Magnetic fields more closely approximating the Earth's field today have also been used to stimulate organisms. In some instances the magnetic field seemingly applied directional clues (Barnwell and Brown, p. 275, p277, Pittman).

Where steady magnetism, regardless of strength, seems to be beneficial (Hays), magnetic variability seems to induce pathological effects, even in modern humans; coronary arrest correlates strongly with extended intervals of disturbed magnetism (Malin), psychiatric hospital admissions correlate less strongly (Friedman *et al.*). Sudden biological extinction has been linked to periods of magnetic confusion in the paleontological record

---

lower energy radiation, which is present in greater amount.

[65] Presuming that the same processes took one gigayear in the primitive environment of Earth, as is postulated by currently accepted theories.

[66] Recently a series of papers in *Nature* and elsewhere, also the book *Lifecloud*, authored by Hoyle, Wickramasinghe, N. C. and others, has considered the possibility of life, now on Earth, having originated from simple molecules, which populate the cold interstellar gas clouds.

(Whyte, p681). Such periods, in our view, would be more likely produced by cosmic large body encounters that would inject magnetic disturbances along with other disastrous effects upon the biosphere.

To summarize, in regard to the time available for the origin and development of species, the Solaria (SB) model is 2000 times less "effective" than the Evolutionary (E) model. With respect to the volume of the life-generating region, the SB model is six million times more effective. Considering the energy density, SB is five hundred thousand times more effective following the establishment of the binary arc. Actually, before its establishment, the nova phase, lasting for months, would have organized the Solaria Binaria system to the equivalent stage of two billion years (2 aeons) of conventionally ascribed Earth history. Hence the SB model, assessing energy density, would well exceed by a millionfold the E model. Since mutagens work upon mutable forms, and branching of species is an exponential concept, the effectiveness of Solaria Binaria in quantavoluting life is multiplied again by the volume of the life-generating region. So, even on a short time schedule, Solaria Binaria appears to be millions of times more capable of producing the species of today.

Still, even this might not be enough to originate and develop the species. The first stages of life are of such low probability, and the later stages of higher but still low probability, that a "guiding factor in life development " must yet be sought. For example, an average protein is formed of a chain of about one hundred amino acids. To quote a creationist: "If all the stars in the Universe had ten earths, and if all of the earths had oceans of 'amino-acid soup', and if all the amino-acids linked up (randomly) in chains 100 acids long every second for the entire history of the Universe, even then the chance occurrence of a given very simple protein [10^{-130}] would be inconceivably remote" (Stengler, p. 16).[67] And the building of a protein is only one of many complex arrangements adding up to life as we know it.[68]

The model of Solaria Binaria might only serve to supersede conventional theory of the evolutionary process, and not to discount it and provide an alternative positive theory, were it not for its electrical features. Life begins by microscopically mimicking its gigantic progenitor, the sac. It has no choice. Every atom, in endeavoring to hold its electrons or gain

---

[67] Insertions ours, taken elsewhere from Stengler's paper.

[68] The variety of propagating forms in the plenum probably extended beyond the mainstream of life. Groups of biological polymers separate spontaneously into coacervatives, small droplets of diameters to 500 micrometers. Where they can metabolize, coacervatives are stable, and can grow and divide. These active droplets are regarded as analogues, not ancestors, of cells (Dickerson).

others, seeks to surround itself with the smallest and densest complete electrical perimeter possible. This is usually an octet of electrons. Whenever necessary, atoms aggregate into molecules where a compromise sharing of electrons will lead to a higher density electrical perimeter.[69] From here the molecules proceed to more complicated systems that ultimately come alive.

The concept of life therefore is an extension of the concept of the "cavity" with which our book began. Life is a way of gaining, hoarding, and begrudgingly doling out electricity. In countless numbers organic molecules determinedly build themselves micro-sacs of chemicals in reaction to electric gradients, capture raw materials, manufacture compounds within the sacs, fire themselves with ever accumulating electric charge, until, incapable of continuing this process without bursting their sacs, they force out unused parts. Usually these are excreta. In critical cases, they are replications of themselves - if not exactly so, then in fundamentally similar ways. No cell divides itself in mirror like fashion, uniformly, in the beginning. But every deviant is a candidate for the first exact mitosis.

The step from excreta to exact reproduction is critical. The sac of organic electrical activity is not "intelligent" except by human prejudices, *ex post facto*. But the sac can most efficiently - effectively and reliably - excrete if it separates its ingredients on the binary principle of "one for you and one for me". Least change, least imbalance, and therefore longer life ensure if the sac polarizes uniformly prior to excretion, setting half of its contents opposite the other half and splitting itself down the middle, closing the gap at the instant of its division. Excretion becomes reproduction.

Sacs that thus form cells which divide offer more chances of survival and conquest of space by numbers than sacs that either hold their accretions until they burst or bifurcate inequitably from an electrical standpoint, thereupon having to internally reorganize their electrical accommodation upon every mitosis. One notes the terrific speed with which life can develop and reproduce under rules of uniform mitosis. Within a few thousand years the plenum might be filled with such cells. Indeed, perhaps large areas were filled with them.

One is not permitted logically to adjudge life as superior to rocks, which have their own form of durability. The biosphere today is a tiny fraction of the rock masses and space of the Universe. As an offshoot of universal

---

[69] Molecules often assume distorted shapes to achieve this compromise. If a spherical arrangement is possible, it is preferred to all others.

change it has a special interest and importance in the perspective of the human mind. Life has a special mode of material extension which, after all, could fill the Universe promptly under proper conditions, and this is a constant challenge to the entropic concept of the Universe.[70]

Life's arrangement of electrical signals is perhaps its chief embedded characteristic. "Electrical potentials occur in all cells studies thus far, although their biological importance is recognized in only a few cases" ("Cell and Cell Division", *Ency. Brit.*, 1974, Macro. vol. 3, p. 1050). The surface of cells is negatively charged. The cell membranes are 6 to 10 nanometers thick and are highly resistant electrically (from 1,000 to 10,000 ohm/cm 2). They produce voltage gradients which drive the biological functions (as noted ahead) and produce a cell interior that is more highly negatively charged than the surface layer of the cell. That cells are so electrically arranged is understandable when one considers charged cells in a charged universe. In metaphorical language, the overall picture of the cell, and the image of the primordial cell, then, is one in which a peculiar combination of chemical compounds survives by erecting an electrical screen to admit nutrients and to repel destructive invaders, then organize its internal components to sustain itself and to resist random escape from the community.

Several varieties of cell growth and transformation are observable. The "main" type of self-duplication ensues as a permissible, organized, collective escape, or excretion, providing for the maintenance of a complete defense system. Cell division would operate by an electrical signal system. The members are an electric grid (as in a vacuum tube), and acts as a gatekeeper among the elements in and surrounding the cell and during mitosis.

Cells make macro-molecules, including genetic molecules, which do not exist elsewhere in nature and are not allowed exit through the cell membrane. Inasmuch as macro-molecules are concentrators of electricity, this synthesis permits the cell to sustain longer than otherwise would be possible its quest for additional electrical charge. The cell thus builds a higher concentration of charge than is available elsewhere in the plenum mixture. This process is the essence of metabolism.

Metabolism concentrates electricity in the macro-molecules, thus depleting of its nutrients the medium trapped in the cell. (The analogies of

---

[70] The Universe is supposedly increasing its entropy with time, that is to say, the parts of the Universe become ever more disordered. Living systems represent increased order because of their internal organization.

cell as sac and of nutritive medium as plenum are close and possibly homologous.) The cell responds by excretion of water, ions and gases (by-products) and ingestion of electron-rich nutrients.

Strain is imposed upon the cell membrane, for it must both contain the increased material and at the same time defend the cell against penetration by electron-deficient atoms and molecules. The membrane signals the cell nucleus concerning an imminent site of charge deficiency and leaking. Then the genetic macro-molecules of the cell, which are the only ones capable of dividing themselves more or less equally, and have been so doing since their last episode of cell division, respond to the signal of impending disaster by completing their synthesis, and by lining up on the two sides of a perimeter membrane that is being electrically trenched through the nucleus at the future site of fission. Actually, the division line-up is provoked by an electric polarization of opposed centrioles, each representing a focus of peak negative charge on the edge of the nucleus.

Midway between the two centrioles, the newly forming perimeter constitutes an electron-poor trench. Following the genetic molecules, the other materials of the cell are drawn electrically to flow in equal amounts to either side of the perimeter-to-be, pursuing the two centrioles. By contrast the cellular material that is to constitute the cell wall itself flows into the trench from both sides. Thus, without breeching its old perimeter membrane, the cell has doubled its surface and has divided. Electrical forces move the two new cells apart. Never are two cell membranes in contact even in a densely packed tissue. Some 15-20 nanometers of intercellular space, filled with a sugary fluid, separate them.

From the self-reproducing cell to the hominid of a few thousand years ago requires passing by many landmarks in the organization of life.

Close to the solar nova and birth of Solaria Binaria at the beginning of the Period of Radiant Genesis, one may position groups of critical developments: the provision from solar debris of chemicals and transmutations in the plenum; and prebiotic organic molecules (amino acids, sugars, nitrogen bases, plus other compounds).

Cell membranes, left-handed symmetry of organic macro-molecules,[71]

---

[71] The origin of one-handed symmetry was probably in the magnetic field (see Edwards et al.). Committed to spiralling into right-handed helices, the DNA molecule and all of the molecules with which it transacts profit from the design, for they thus attain denser molecular packing, producing greater electric stability. The tightest-packed helix is the alpha right-handed (screw) helix - here each turn of the coil incorporates 3.6 to 3.7 amino-acid units. This form of the helix has no open spaces in the center; further, all amino-acid structures are exposed on the surface of the helix (Mazur and Harrow).

proto-enzymes, porphyrins and => *nucleotides* - these developments would readily follow. The cell probably took in the latter three constituents after proto-proteins had been formed independently in the plenum. Some cells, instead of dying, began to engage in mitosis, whereupon self-duplication, as described here, would soon follow.

Large cells would ingest small cells, or form around them, performing two types of action: digestion, the beginning of animal behavior,[72] with the breakdown of the electrical defenses of the smaller cells, and in other cases the formation of cell colonies using the membrane of the host cell as a super-membrane or skin of the smaller internal cell or cells. Large cell colonies would float in the magnetic tube and, later on, settle upon solid bodies.

From the development of the cell, the mode of basic change in life forms ever thereafter can be surmised. Time after time it happens that some portion of the excreta of the organism is retained within the sac of the colony and supplied with the coded electrical signals that connect with the master genetic material so that its descendant in the next generation can draw upon its experience and existence. The developing special organ excretes within the organism and returns signals to make demands, denote satiety and share directiveness in the behavior of the full organism.

For example, the eye is always close to the mouth. The photo-receptive organ that perceives food chances is close to the sac opening that can employ opportunities for ingestion. The organism as a whole is, as it always has been, ready and eager to accept charge-bearing contributors which allow it to increase its density. (It rejects cations for this reason). It permits and then becomes dependent upon the vision, with the genetic material duly recording and perforce returning in the form of instructions the interrelated, combined signals of the eye-mouth.

The genes do not "know" that they are building an eye to go with the mouth; nevertheless, they do so with despatch, as they eagerly accept extensions of all such special organs in the Period of Radiant Genesis; for the environment has a plenitude of electron-rich chemicals, a state of affairs that does not persist beyond the first half-million years of Solaria Binaria. In more modern times, the cell (and hence the organism as a whole) is more hard-pressed to find energy-rich molecules and in the very stress to obtain nutrients it has bureaucratized itself so to speak, and is

---

[72] We see certain bacterial and plant behavior in photosynthesis as a concurrent development, supplementing an animal diet with the capturing of a chlorophyll (pigment) molecule, precursor of the protein, which was useful in the internal manufacture of foodstuff.

hence even less equipped to obtain them. In the modern electrified environment, vital processes take much longer.

The plenum of Solaria Binaria was the creator, cradle, and mutagen of life. The broad sculptures of plants and animals were completed during the first half of its existence. If fossils represent the basic variety of life, the phyla and the orders came into being then. No new general forms have originated in recent times (Brough). Despite great waves of extinction, slightly over one million living species are named today. The fossil record should show millions of ancestral species to provide the present number, but in fact shows only about one hundred thousand species. This contrast has excited comment: why were large changes peculiar to early existence; why were small changes more common in recent times *(ibid.)*?

Set up in this manner, the questions seem to accept answers from Solaria Binaria theory. The plenum promoted creation initially, as would be expected, promoted it less when the binary was stabilized, and became quite destructive and conservative as it exponentially decayed and collapsed. The agents of these change may be identified. The first period provided an immense number of prototypes and access to abundant nutrients, so testing their viability (Ayala). The second period provided a stable environment of abundant nutrients but an end to the easy method of forming combinations. Further, the more distinctive and specialized the species, the less likely its electrical transformation would eventuate in new designs of life.

In the final period, environmental disasters extinguished many species, but also promoted very many, already genetically deviant individuals to the status of families, genera and species.

To acknowledge that a great many of these lesser, less creative designs have emerged in the later history of Solaria Binaria requires a theory of genetic realization. The genetic material can carry far more instructions for the construction and behavior of any organism than are required at any given time (Ayala). Under lower (but higher than present) solar system quantavolutionary conditions, suppressed instructions can be triggered. It is conceivable that every living species carries in its genetic code instructions for metamorphosis (monsterism). Cosmic rays, nuclear explosions, radiation fall-out meteoroids, electromagnetic typhoons, encounters of Earth with large bodies (comet, meteoroid), viral epidemics, and "silent" significant changes in electrical discharges within Solaria Binaria and the Solar System may be the means of suddenly extinguishing some genetic instructions and releasing others, quantavoluting a species into a similar but substantially modified species that is altered anatomically, physiologically and behaviorally.

Success has not attended the search for transitional forms that bridge the "gap" of development from one species to another under conventional Darwinian theory. It may be maintained that transitional forms, such as reptiles with half developed wings or hominids that spoke but poorly, never existed (Rodabaugh, p. 119). All orders of mammals appear with their "basic ordinal characters" (Simpson, 1944, p. 106). Many of the plant species, it is believed, are replicas of other species (=> *polyploids*), differing almost entirely in size alone, with the physiology and behavior appropriate to giantism and dwarfism.[73] That the horse, a favorite instance of evolution since Lyell, has evolved its peculiar configuration by means other than genetic realization seems unlikely. The millions of years authorized to complete this series of changes (among others) are unnecessary and probably even insufficient unless supported by a theory of genetic realization, a position that has forced its way into contemporary evolutionary thought to evade the constraints of ever greater stretches of time and of evolution by random mutation under uniform Solar system conditions.

The problems of explanation that remain are historical and technical, inasmuch as a common electrical process is followed in all biological changes. The applications of the process - to change marine animals into amphibians, reptilian types into mammals, one animal into another with all the anatomical, physiological and behavioral changes involved - occur according to a simple set of principles. Nor are these adaptation, nor survival of the fittest, nor random successful experimentation with mutations, all of which are minor aspects of quantavolutionary change. Rather, electrical claims are provoked by opportunities, encounters and transactions, and organize themselves into genetic storage and release.

Evidence from the surface of the smaller remaining planets shows total devastation and almost total loss of atmosphere. On Mars, where some atmosphere remains, no biological residues survived (Horowitz, p. 55). The Martian surface was found to be so deficient in organic material that a mechanism for their removal is being sought. The inner Solar System is now sterile, excepting Earth's biosphere, which thrives.

A final short period follows the period of evolution; it is an epoch of explosive quantavolution that comes down to the present. It witnesses catastrophes of life forms, quantavolution through genetic realization, and

---

[73] One-quarter of the flowering plants may be polyploid species. Some vertebrates are polyploids as well (Tinkle).

the rise of *Homo sapiens*. On the physical side, it carries the record of the destruction of Solaria Binaria and the advent of the Solar System. Though short, this period contains the full human experience. Its story forms the second part of this book.

PART TWO:

DESTRUCTION OF THE SOLAR BINARY

CHAPTER TEN
INSTABILITY OF SUPER URANUS

The first part of our narrative was given over largely to the origin and progress of Solaria Binaria up to the beginnings of a fatal instability. Almost from the beginning life burgeoned and flourished in the plenum, and subsequently on Earth and other planets. Celestial objects were not then visible from the Earth because the plenum was too dense to let light pass directly from the binary stars to the planets. Hence mankind originated its physical being almost entirely in a world where murky grey skies softened the light through a misty air. His psychic being and intelligence, by contrast, were formed in most important regards during the degeneration of the binary system.

As best we can locate this turning-point of humanity, it happened during the pre-nova instabilities of Super Uranus. Already mentioned is evidence that early humans had intimations of a primordial plenum and an electrical fire. More extensive evidence correlates human observers with the expectable, inferable, behavior of Solaria Binaria as it would begin to collapse. The first human observations have to do with a solid heaven that began to separate from Earth and fell apart.

A number of peoples claim that the primeval chaos was present before the creation. It was a plenum - dark (compared with what followed), uniform, dense, and housed a Demiurge who had not yet acted and a world of things and beings that were potentially activatable. The Ngadin Dayak, a people of Borneo, insist that "at the beginning, the cosmic totality was undivided in the mouth of the coiled water-snake", possibly referring both to the togetherness of the chaos and the omnipresence of the electrical axis mundi. In the Hindu Vedas, Dyaus-Pitr (Dyava), "Father Sky", can be identified with the age of first man and an unbroken plenum. He married Prthivi, Earth. The world was dark and asleep, "says" Manu (a Hindu Noah) until the Great Demiurge "appeared to scatter the shades of darkness".

Coelus, or "Heaven", was the most ancient Latin god of the sky. His name means "covering". Ouranos was "Heaven" and the first god of Greek legend; this Heaven was at first a calm and settled person, married to Mother Earth, Gaea. The Chinese legend pictures Heaven as T'ien, at first a marble-like ceiling, unbroken. According to the Iroquois of north Eastern America, the Chief of Heaven was persuaded into marriage by "Fertile Earth" (Awenhai) and impregnated her by his breathing. The Hebrew Book of Genesis, a creative compilation, probably by Moses, of earlier legends, describes in its opening verses the Demiurge brooding over the combined celestial and earthly universe; "The Earth was a formless void, there was darkness over the deep, and God's spirit hovered over the water." As the editors of the Jerusalem Bible comment, most of these images are intended to describe how being may be created from Nothing.

All religions, says Eliade (1954, p. 4) go back to the earliest times, *illud tempus* (" That Time") when the world was born and the initial creative happenings occurred in all aspects of existence. Ever thereafter, the practices and rules of the religion are obsessed with repeating the events of those days. It is obvious that all peoples look upon this epoch, *illud tempus*, as a highly volatile quantavolutionary period, full of stresses and inventions. There is no uniformitarian or mild *illud tempus*.

In many places, a theory of the Cosmic Egg is used in connection with the earliest god, who is Heaven; it explains how God and the World were born. Thus the Hindus asserted that a seed was laid and became the Golden Egg. The Cosmic Egg is often said to have existed from an age before it revealed itself. We construe from this that the earliest humans are present on Earth as the troubles of Solaria Binaria heighten and that they were newly human for a short time before Super Uranus, that is, the Cosmic Egg, appeared.

Other widespread types of creation legends are usually conformable to the Cosmic Egg myth or do not contradict it (Long, 1963), and often occur within the legendary corpus of the same culture: as examples, in the Greek legends of Hesiod and the Cosmic Egg myths of Orphism, in China with two variant stories about P'an Ku, or with the Dogon of the southern Sahara, who put creative twins within the Cosmic Egg. Some type of creative urge is antecedent to the Egg, usually a supreme sky deity. The Egg can take related forms: a shell, a seed etc. Fulminous powers that break out of the Earth to assume living forms are a type of creation widely believed; usually an external force inseminates or provokes the Earth, as in the Egyptian image of Nut and Geb, Heaven copulating with Earth. Finally, parents representing Sky and Earth are experienced as separating, allowing life to flourish. Earth divers, yet another type of creator, are deities who often are commanded by a supreme deity to plunge beneath the primordial waters of chaos and emerge with Earth, which becomes the site of life. This cosmic image can be a metaphor of the same events in the breaking of the Cosmic Egg and the separation of the Sky and the Earth.

We assume that primordial observations gave rise to all of these legends. As part of the continuing emergence of life and intelligence from the chaos befalling Solaria Binaria, the Cosmic Egg myth is imagery for the vision of the first great "sun", Super Uranus, as it emerges from the Heavenly gloom. It emerges shortly after human self-awareness, and in a time of troubles for mankind. This Uranian Age was, both in legend and in our astronomical theory, a time of disturbances.

As is typical of evolving binary systems, the principals, here Sun and Super Uranus, move apart with time. During the phase when a strong electrical current flowed between the two stars the components remained relatively close together, while the whole system charged negatively in transaction with the galaxy (Figure 21). This charging drives the Sun and Super Uranus apart so that the current flowing between them weakens and from time to time falters; the two become more isolated electrically within the ever-diluting gases of the plenum. The importance of the electrical axis of the binary diminishes. The universality of binary recession can be documented by Russell's 1927 data, where star class is correlated with binary period. Only Bruce (1944, p. 13) has connected binary age with separation of the components.

While the two bodies, the Sun and Super Uranus, were transacting vigorously, they were quite luminous, though not sufficiently bright to be perceived as celestial bodies. As the arc between them began to falter, they and the arc remained luminous, but the latter less so than before.

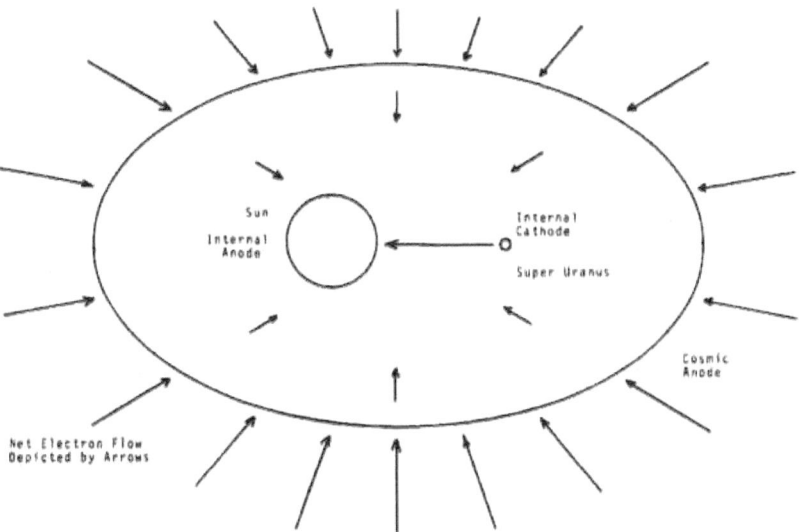

Figure 21
Transaction Between Solaria Binaria and the Cosmos : Dense Plenum Phase.

Originally Solaria Binaria transacted with the Cosmos as a unit. Electrons flowed from the Galaxy into the sac of the binary, liberating energy to the gases along the perimeter of the plenum. Charges so delivered to the binary as a whole were subsequently redistributed among the parts of the enveloped binary, producing secondary energy releases as they impinged upon some particular component (star or planet) within the dense plenum.

Though the central arc was sputtering, the surrounding gases in the magnetic tube sustained an afterglow and so were not always extinguished between discharges. But, as the arc decayed further, the discharges became less frequent, so that eventually even a long afterglow could not maintain continuous luminosity throughout the magnetic tube. At times the sky briefly darkened. This was the first light and darkness experienced by humans, who may then have deduced the concept of contraries, good-bad, yin-yang, or light-darkness, the basis for religious dualism and human thinking processes.

Relative time may have been invented in the period of Super Uranian instability. If the arc pulsed regularly, the earliest humans would have responded to it, first subliminally and later consciously. Possibly, when the glow of the arc darkened and lightened perceptibly and in rhythm, a notion of periodicity would be imparted to humans: they would have a clock. The experience of the first abrupt darkness would be terrible, the unexpected gloom of even a minute, provoking fears of a shutdown of light. If the pulsing were regular, but interruptions occurred, terror would ensue with the interruptions. Frenzies of fear attending eclipses, historically and recently (Corliss), may be traceable to primordial experiences with a degenerating axis. The birth of Super Uranus, emerging from the plenum sky, would be both terrifying and reassuring. Graeco-Roman mythology pictures the god Uranus as gloomy and enshrouded (de Grazia, 1981). New measures of time and space might be calculated, a reliable presence was granted humans, and even the ultimate terror of a turn-off of electrical axis activity could be tolerable if Super Uranus remained visible.

It is possible that through this period the electric discharge was converting from one emitting light to a non-optical, or dark, discharge. Thus the absence of light is not a synonym for the absence of electric flow, only a change in the gases' reaction to the flow. Nevertheless, pauses in the glow were becoming longer with time and the flow more erratic in its intensity.

Whereas before, the two stars transacted internally to produce the arc, while the opaque plenum transacted at its perimeter with the Cosmos, now each star transacted separately with the galaxy through the thinning plenum. This new galactic connection could occur because the plenum density had fallen as it expanded; both bodies were still far from electric equilibrium with their galactic environment (Figure 22).

Thereupon, Solaria Binaria would be observed to be a semi-detached binary star system (Note D) from the vantage point of another star system. Not surprisingly, the gas density detected in such binaries is at a level of that plenum density that would suffice to let the principals be seen from the Earth's presumed location during that era.[74]

An important feature in semi-detached binaries is the flow of material from one of the principals to the other. Usually gas flows directly between the principals; however, in some cases the flow is deflected or is directed around the recipient star (Batten, 1973a, p. 8). Gas expanding from the star is seen in some systems.

---

[74] About 1.6×10^{-6} kilograms of gas per cubic meter, or 10 atoms per cubic centimeter.

Figure 22
Solaria Binaria as the Plenum Thins and the stars Separate.

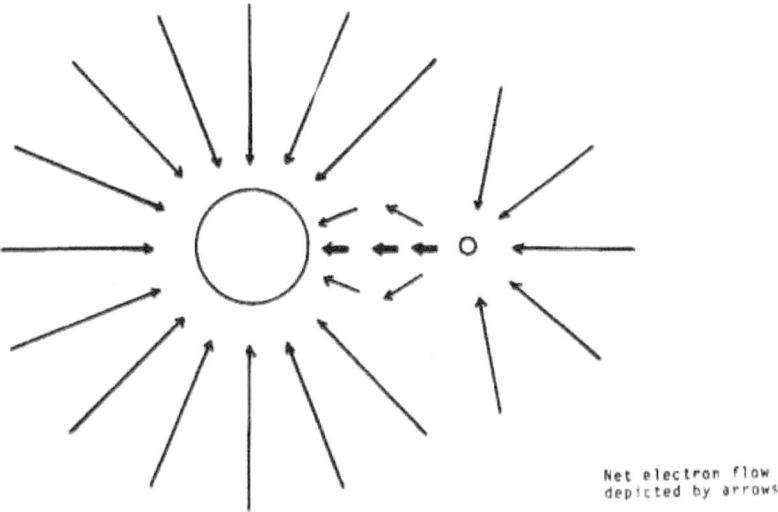

Net electron flow depicted by arrows

Late in the development of Solaria Binaria the gases of the plenum had been thinned to such an extent that the transaction between the Cosmos and the binary ceased to liberate the major part of its energy at the binary's perimeter. Thereafter the cosmic transaction deposited energy individually at the two stars. The inter-star transaction (the electric arc) continued for some time after each of the stars attained a separate connection with the Galaxy.

Wyse showed that emission lines are often observed in close-binary systems. Emission by hydrogen, helium and singly ionized calcium is common. Mass transfers amounting to 10^{-8} to 10^{-5} Sun per year have been proposed in such systems (Koch, p. 90). In some cases, lengthening of the period of the binary has been ascribed to mass loss from the system (Nather and Warner). In our view the flow of mass contributes towards separating the two stars; nevertheless most of the separation occurs because of the electric charging of both stars through transaction with the Cosmos. We also see no way for gas to escape except as stellar wind.

In certain close-binary systems, like nova AO535+26 in Taurus (Coe *et al.*), intense X-ray emission is noted as gas flows onto one of the stars

(Wickramasinghe and Bessell). The discovery that most galactic X-ray sources represent close-binary systems and that in some cases a flow of ionized gases occurs between the principals in the presence of (inferred)[75] magnetic fields is an important finding (Kraft): we concur. The presence of electrical potential difference between the two stars makes X-ray emission understandable. Detection of cosmic X-ray sources implies that electric, not magnetic, behavior, is being observed. We infer that Solaria Binaria was an X-ray emitting binary at this stage. Enough of these X-rays were penetrating to the planets to cause quantavolutions in the biosphere before the eyes of humans, possibly contributing to the age-old beliefs of humans in metamorphosis of living things.

Various scholars have maintained that all peoples have possessed religious beliefs from their earliest origins, that these beliefs centered upon a single "Heavenly Father" as a type of monotheism, and that this father God became indistinct after the first ages, was lost, was forgotten, and/or was indifferent. This is true of Dyaus in India, of Ouranos in Greece, of Coelus in Latium, and of the "Great Fathers" of the Australian Arandas, for example (Eliade, 1967, pp. 20ff). This Ouranos-type is first the sky and then the materialization of the sky into a sun-like body, whence it disappears and is replaced by a son, a Saturn (Tresman and O'Gheoghan, p. 36).

The Maori of New Zealand have the Demiurge moving from inactivity to increasing activity. And this is a universal subsequent theme. The skies break up. They fall. Humans are much disturbed. Their world changes. The heavenly god moves heavily and destructively. The god becomes various gods, families of gods, and demons. Creation is under way, rarely in one phase, but continuously, over a long time - thousands of years, we think - before arriving at what is recognizably the modern Solar System.

Hesiod's version of the Greek creation myth has Ouranos or Heaven squeezing down upon Mother Earth, oppressing her until she cries out in agony. We interpret this suffocation of Gaea as an increasingly disturbed atmosphere, with many extinctions and quantavolutions in the biosphere. The mechanism usually termed "natural selection" operates rapidly, under extreme environmental pressures. The "fittest" which survive are often accidents of isolation, or species that can draw upon luckily beneficial reverse or recessed genetic capabilities, as well as groups with now to be proven superiority in food-finding and breeding under difficult conditions.

Ouranos goes increasingly mad, taking up his children and hurling them

---

[75] Reservation ours. Conveniently, such fields are not generally detectable (Batten 1973a)

beneath the Earth, Gaea in desperation urges her brood to revolt against the Father. We interpret his sons plunging to Earth as a bombardment by heavy meteorites, released into the plenum by the "unsettled" Super Uranus and encountering the Earth.

To the pre-nova turbulence of gases and bombardment, a duration of 1000 years may be assigned before the human creation (which will be related in Chapter Twelve) is connected with the turbulence; subsequently another 2000 years is assigned before the climactic nova of Super Uranus.

In China, P'an Ku, a creator god, began pounding on T'ien, breaking large chunks off with his hammer and chisel until the skies showed through. The Dayak of Borneo report that two mountains arose and clashed, with the first features of earth and sky emerging into existence from their explosive contest. The mountains are revealed as two creator gods, Mahatala and his parahedra, Putir, who then continue to create.

Turning to Hindu sources, Dyaus is replaced by a struggle of two types of heavenly powers, one good and the other evil, one led by Varuna, the other by Vitra. The good powers were termed Adityas; the bad dragon-like demons were called the Vitras.

Removed from the protective blanket of the plenum, which heretofore had isolated Super Uranus from the Cosmos, this sun-like body became directly subject to variations in the electrical environment through which it was travelling (Chapter Three); a new variability of the surrounding plenum's electrification was produced by the sputtering arc.

At this stage, Solaria Binaria was transmogrified and might have resembled the cataclysmic variable stars, a group of close binaries. Here the primary is sub-luminous and its companion is often a dwarf red star. The diminished luminosity of the stars begins as the components readjust from internal transaction to galactic transaction. We think that, in transition, Solaria Binaria, now a low luminosity system, entered an eruptive phase.

That Super Uranus was the erupter in these first noted celestial events, and not the Sun, is confirmed by the evidence that the ancients did not regard the Sun as a powerful sky god. As de Grazia has noted elsewhere (1981, p. 258), "the regularity of the Sun, once it appeared in the skies, worked against its becoming a great God".

In ancient writings the planet gods sometimes altered the motion of the Sun and the stars, but never the converse. Occasionally, as part of a catastrophe, the Sun would go on strike, bringing up darkness. Velikovsky (1950, pp. 300ff) thinks that Macrobius in the fourth century may have been mainly responsible for the erroneous personification of many sky gods as the Sun. We can say that at least he represented a trend of ideas, which Jacquetta Hawkes has confirmed (de Grazia, 1981, p. 259). Closer to our time, Max Muller's extensive work on primordial religions has imprinted this error in the minds of most scholars.

The outer layers of Super Uranus and its => *space-charge sheath* were the first places to react to instability. At intervals, a shell of material expanded explosively away from Super Uranus. To the outside observer this small star had become a nova of low intensity. Weak outbursts are not uncommon in under-luminous close-binary systems.

Some close binaries contain dwarf-nova stars, for example, SS Cygni. It is possible, sometimes, to see a hot spot where gas flows from one of the principals onto the atmosphere of the other (Cowley *et al.*, 1977, p. 471; Hesser *et al.*). Dwarf novae also exhibit flickering, which usually disappears if one component eclipses the other. The flickering, which is especially intense in the case of Z Chamaeleontis, is attributed to the hot spot (Mitton, pp. 84ff).

On the other hand this flickering may be a variation of the current onto the photosphere of the stars as the system adjusts its mode of transaction from that in Figure 21 to the one shown in Figure 22.[76]

Many stellar binaries involve components which have perplexed astronomers, because, according to the criteria of classification, one star is very old while its companion is quite young (see Kopal, 1959). Usually these pairs are closely orbiting as we propose was Solaria Binaria. Such pairs, with discrepant evolutionary ages, are thought to be systems in which one component has passed through the nova stage, some indeed being recurrent novae. Krzeminski believes that in U Geminorum the irregular flow of matter from the red companion triggers recurrent nova eruptions on the white primary (see also Aller, p. 603). The primary star in such systems is usually classified as a white dwarf star (Glasby, p. 61). A cycle amplitude relationship has been established linking the intensity of the recurrent nova flare-ups to the time between recurrences (Kukarin and Parenago). The larger the flare-up, the longer the recovery. For the largest flare-ups, recovery time exceeds the period of observation; here periodicity is implied rather than established. The regular recurrence may be a discharge effect. The transaction between the star(s) and the Galaxy may slow down periodically due to space-charge fouling of the discharge channel. The discharge then diminishes, which allows the interfering

---

[76] Juergens (1977d) notes that similar current variations exist in the solar photosphere.

space-charge to dissipate. A new breakdown can now occur, leading to another flare-up.

Alternatively, it may be that, at this time, Solaria Binaria moved into a region of the galaxy in which the cosmic electrical pressure was diminished (see behind, Chapters Three and Four). The binary, and especially Super Uranus, as the smaller, highly electrified part, could teeter on the verge of serious internal instability. This condition, which may have persisted over about three thousand years, would have proved disastrous for Super Uranus and eventually altered for all time life on nearby planets, including the Earth. Milton (1979, p. 74) has postulated that the Sun today remains stable relative to the cosmos surrounding it on a moment-to-moment basis; small solar inconsistencies have been noted over the historic period (Eddy *et al.*, pp8-9; Clark *et al.* 1979). Even the ultimate instability, the nova eruption, is not forbidden.

Super Uranus can have erupted many times. The shock of its recurrent explosions propagated through the plenum, damaging the planets nestled within it and electrically thinning it further. Not all of the ejecta was gaseous. Fragments circulated within the system for a time, encountering explosively other bodies differently charged. Some fragments fell back upon Super Uranus, which was diminishing in brightness and maybe also in size because of its outbursts.

The rude disturbance of the hitherto peaceful atmosphere of the Earth was noted fearfully by the rapidly developing human culture that was spreading throughout the World. Men perceived the heavens to be alive and exercising a control over earthly affairs. Heaven both inflamed and frustrated man's desire - which seized him in the course of his very creation - to control himself and his environment.

CHAPTER ELEVEN

ASTROBLEMES OF THE EARTH

The first experience of Super Uranian instability on Earth would be a quick succession of light and darkening and a relatively more pronounced illumination from the South (Sun) and the electrical arc. The dark would come from the expulsion of dust and debris down the sac towards the Sun (see Figure 23). Pandemonium would be let loose and frightful specters abound as fragments would rip through the plenum and encounter Earth.

Instability of Super Uranus periodically expelled from that body a halo of debris whose nature depended upon the intensity of the particular outburst. It is conceivable that the process could persist over several millennia with frequent small eruptions occurring at intervals similar to an active volcano or to a recurrent nova (Chapter Ten). Mild outbursts might only cause ejection of superficial material gases and fine solids. Violent ejections could send massive chunks of solid material away from the star. Because the binary is nestled in the cavity, the ejecta does not escape the system. However, its fate is dependent upon its electrical state and the direction of ejection.

In its outbursts Super Uranus mimicked, but with diminished intensity, the nova eruption which the Super Sun underwent early one million years earlier. Electrical instability between the skin of Super Uranus and its interior, probably produced by the transition between one mode of transaction and another (Chapter ten), led to explosive ejection, in all directions, of layers of the star. Much of it was captured by and funneled down the magnetic tube. Its penetration towards the Sun was governed by its inertia and charge (see Note C). This material, possessing greater charge density than other parts of the binary system, caused havoc as the pieces (atoms to irruptives) encountered the plenum gases and the planetary bodies.

The electrical, meteoritic, and gaseous disorders attendant upon the initial instability of Super Uranus are largely deduced from the dynamic model of the collapse of Solaria Binaria. Direct proof of the falls associated with system derangements extending over a period of perhaps three thousand years is lacking. In an extreme case it may be postulated that most of the damage of an extraterrestrial meteoritic character belongs to this period, as opposed to damage inflicted by planetary size bodies to be discussed later.

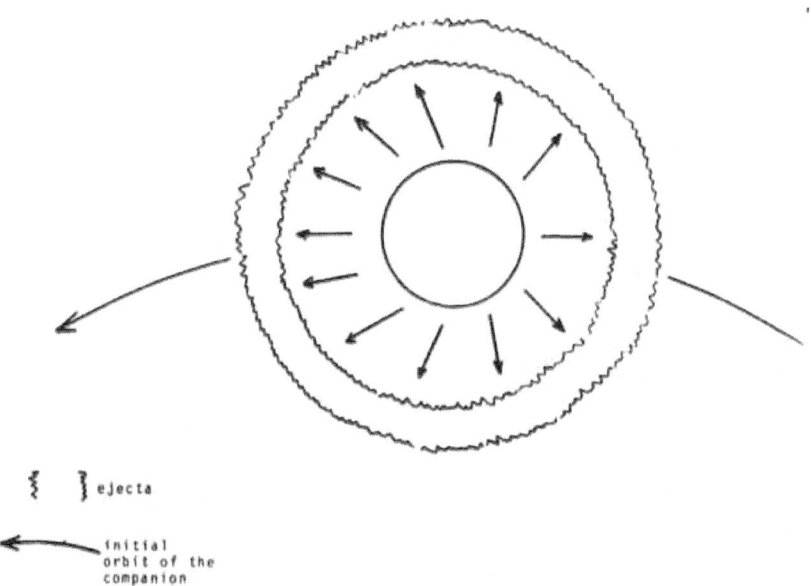

Figure 23
Explosive Eruption from Super Uranus.

At the period when the galactic transaction to Solaria Binaria was shifting from the gases of the outer plenum to the gases closely about the two stars, electrical instability developed within Super Uranus. This instability caused Super Uranus to shed explosively material and gases from its body. Much of the ejecta became trapped in and funneled down the magnetic tube, bombarding the planetary components of Solaria Binaria.

Probably impacts were rare during the period of stability following the first accretional stage of the Earth. Evidence for this in rocks and depressions would have been metamorphosed, granitized or erased under sedimentary aggregation and erosion.

We suggest that most extraterrestrial deformations of the Earth's surface would then have occurred at the end of the stable period, that is, from fourteen thousand years before the present onwards, during the period of Super Uranus instability. The lunar episode, to be discussed in Chapter Thirteen, would have provided most of the remaining meteoritic features, or astroblemes. Here the material itself would have been mostly a fall-back and possibly identifiable as Earth-crustal material by physical and chemical techniques if its nature would not be later modified to conform with Earth. Subsequent disastrous showers of meteoroids, as we shall explain, would have been experienced in Apollo and Venusian times, that is, around 5000 and 3500 years ago.

Lately the term "astrobleme", meaning "star-wound" in Greek, has come into scientific use along with the renewed interest in things coming out of space. Generally it refers to detectable craters dug, supposedly, by meteorite falls. Here, our discussion of astroblemes includes a whole class of effects of extraterrestrial transaction with the Earth's surface: "meteoritic" craters and mounds, irruptives (collisional intrusions that may turn out to be soft-landed meteorites); meteoritic dust; => *barads* and field cobbles; till (consolidated clay and pebbles); metals; ash; waters; ice; vaporites (fall-back of exploded and extremely heated meteoritic and terrestrial material); fulgerites (fused soils of lightning origin, whether terrestrial or extraterrestrial); and biospheric transformation. Controversy and a paucity of identified materials makes this list hypothetical; certainly it is not complete, because extra-terrestrial collisions, small or large, must convey many lost effects. Before long, for example, it will be difficult to detect, even guided by a precise hypothesis, the eighty million trees blasted down in the Tunguska region of Siberia in 1908, probably by a meteoritic air-burst; the animals and few persons killed in this obscure wilderness disaster have long disappeared into dust. Mutated flora has been reported from the spot; such plants would have merged into the plethora of ordinary species if there were not a search party alerted to their possible quantavolution.

Distinguishing among astroblemes of the various episodes 14,000 to 11,000; 11,000 to 10,000; circa 5,000; and circa 3,500 BP; and all others, even though perhaps a minor concern, is probably impossible because of the heterogeneous nature of the Earth's crustal material and the similar processes occurring in each case of a strike.

Legends and history will afford some assistance and could afford more were these now to be reviewed in search of incidents. For some time Australian Caucasians disbelieved the reports of Australian Aborigines that McConnell Bay had suddenly appeared where before there was no water.

Late studies have changed the date of origin of the feature from millions of years BP to a few thousands (Kondratov).

Meteorites were often incorporated into places of worship, as sacred relics of the vitiation of, or a message from, a god. The Temple of Artemis/ Diana at Ephesus in Asia Minor contained a meteorite (Acts 19-35); the image of Diana was reputed to have been sent by the god Jupiter. Velikovsky (1950, p. 289) cites other examples. The best-known surviving meteoritic object of worship is the Black Stone (30 centimeters in diameter) now encased in silver and embedded into a corner of the Kaaba (Ka'bah) in Mecca (Abdul-Rauf, pp. 584ff). A local legend attributes the stone to the Archangel Gabriel who is associated with Venus (Velikovsky, 1950, p. 291). Moslems believe that the stone is the only extant piece of Abraham and Ishmael's House of God (Abdul-Rauf).

The geophysics of crater identification is in its infancy; the very idea of the Earth having suffered extraterrestrial encounters has been resisted until lately (Ninniger). Craters from smaller than seven kilometers to seven hundred times that diameter are discernable under various geological formations at widely separated locations in continental North America and elsewhere (Saul). Ancient meteorite craters may be the source of many circular features of the Earth, but few of such topographical formations have been given more than a superficial look (Norman *et al.*, p. 692). Figure 24 shows an area of broken terrain in Arizona from which Saul's analysis revealed a set of overlapping and eroded astroblemes as shown drawn over the map. Notably, metal and mineral deposits are distributed among these astroblemes, lending support to our suggestion elsewhere in this book that most if not all useful minerals and metals are deposited and produced by quantavoluntionary processes.

Beals and Halliday outline criteria used to identify meteorite crater remnants after erosion, and possibly glaciation, have attacked the exposed circular or oval structure. Critical is the presence of a lens-shaped layer of broken rock under the crater. This is often extremely difficult to reach by drilling. They note that fragments of the meteorite usually are absent; this they attribute to removal by glaciation. However, we maintain that no fragment need have fallen to produce such a crater. A crater produced by the shock from an explosion resembles one produced by material impacting at high energy, both exhibiting phase transitions that produce high density crystals from the resident minerals. Glasses produced by heat also are common in both settings. Craters satisfying Beals and Halliday's criteria result when great electrical discharges reach the surface (Juergens, 1974; 1974/75).

Figure 24
Possible Astroblemes in Arizona.

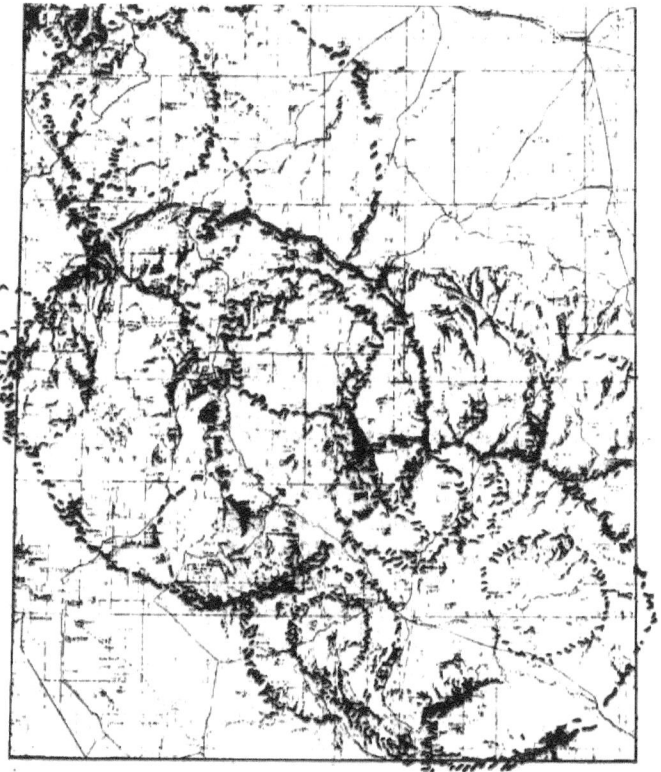

A section of an official relief map showing a portion of Arizona at a scale of 1: 2 000 000. The rectangle encloses the land between 110° and 112° West longitude and 33° and 35° North latitude. The city of Phoenix is located on the west margin of the enclosed area about one quarter of the distance from the bottom to the top corner of the map. The circles representing the remnant astroblemes have been drawn over the map; they are based upon the analysis of Saul. Extensive mineral deposits have been discovered at sites on the rims of these features.

Vsekhsviatskii, speaking about the origin of the Moon's craters, notes that "the magnificent achievements of the Apollo astronauts... leave no doubt that most of the processes affecting the surfaces of the planets were determined by endogenous forces." He favors eruptive genesis, because of the basaltic nature of the ejecta surrounding the Moon's craters. In our opinion, he is incorrect in attributing these eruptions to processes originating within the Moon (and the planets), but is correct in his observation that only local material is present. The same is true for Earth craters. Only rarely do large meteoroids contact the Earth, because of electrical repulsion between the charged Earth and the invader (Figure 25).[77] Some overcome the repulsion and go on to impact (trajectory 3); others do not and deflect back into space. Many meteoroids become unstable and discharge electrically (trajectory 4); the discharge can explode into the Earth's surface, producing a "meteorite" crater, or it can produce an atmospheric shock wave which devastates the surface features. More commonly, a bolide is produced that discharges harmlessly well above the surface; only audible shock-waves reach the surface (trajectory 2).

Then dusty débris or a few small rocky fragments, splintered off the meteoroid, may reach the ground (Milton, 1982). Most meteoroids "burn up" at high altitude (trajectory 1), the smallest of which are noted to decelerate as if repelled by the Earth.[78]

Hughes (1979) commented that certain meteor swarms observed within the Earth's magnetosphere behaved as if they were electrically charged. This conclusion is consistent with the surprising finding that the rate of encounter between Earth and fainter meteors correlates negatively with increased solar and geomagnetic activity (Lindblad). Other charged particles encountering the Earth from directions away from the Sun's show a similar inverse correlation with solar activity, which lends support to the concept of charged meteoroids.

Motion of ejecta, like the motion of the principals, would have been dependent upon the relative charge densities of the transacting pieces. Under stable conditions, the gases and material within the magnetic tube were close to being in electrical equilibrium with the flow along the electric arc. Thus material encountering the Earth should normally have a charge density approximating that of the Earth and would be repelled in encounter. Penetration into the Earth's electric domain (a space much larger than the body of the planet) would be determined by the combination of mechanical inertia and electric attraction/repulsion (see Table 5). Most meteoroids could not reach the Earth's surface before the electric repulsion would reverse their trajectories and fling them away into the plenum; alternatively the electric transaction between the meteoroid and its surroundings would consume the encountering body before it could be repelled.

---

[77] The first documented meteoroid repulsion was made in August, 1972 (Jacchia).
[78] These faint meteors decelerate at rates up to one hundred times greater than that expected for a solid body penetrating the Earth's upper atmosphere (the ballistic meteors).

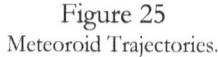

Figure 25
Meteoroid Trajectories.

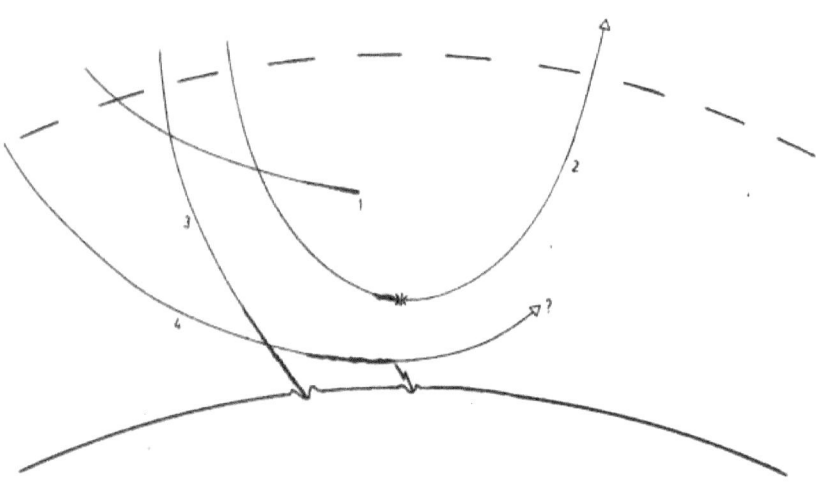

Objects from space that penetrate the Earth's electrosphere and enter its atmosphere transact strongly as they approach the Earth. For bodies larger than a grain of sand a visible trail, a meteor, is produced during the passage through the atmosphere (1). Frequently a meteor will explode harmlessly high in the atmosphere, to produce a bolide (2). A very small fraction of incident meteoroids overcome the electrical repulsion by the Earth and impact with the ground: these are the meteorites both ancient and modern, the majority of which are small and thus can become equilibrated with the Earth's electrical state during their short falls. The largest meteoric pieces can impact explosively (3) or discharge to the ground, damaging the terrain indirectly (4).

This transaction arises because particles of different sizes possessing the same charge density have different electric potentials at their surfaces (see also note C); thus they must transact if in proximity. The larger body has the higher potential and gains charge from the smaller. This heats the meteoroid and may vaporize it. If the potential difference is great enough, lightning-like currents may be induced between the meteoroid and surrounding charges, explosively stabilizing the charge levels; such discharge would be expected only for large meteoroids.

In the disruptive environment, when the binary began to be electrically unstable, large amounts of meteoritic material could encroach upon the Earth's domain, arriving in an electrically inflamed condition (at very different charge density). Some of this material would be strongly attracted towards Earth and could blast explosively into its surface. Even when a near miss occurred, the passage could alter the Earth's protective electrical

Table 3
MODES OF METEOROID ENCOUNTERS

Charge	Inertia		
	Low	Moderate	High
Repulsion	"Faint meteors"	Evasive skip	Air explosion
Neutral	Drift down	Small intrusion	Rafted irruptive
Slight attraction	Ballistic meteor	Fireball	Bolide
Strong attraction	Soft fall	Hard fall	Explosion crater

sheath (as solar wind outbursts, produced by solar flares, do today), great thunderbolts would be generated, and again produce explosions at the surface.

When a tremendous bombardment, or large-body encounter, would occur, most of the matter could not overcome the electrical repulsion of the Earth; but vast sporadic falls from above could dot the Earth's surface. Remnants are found buried under the fallout from later catastrophes (Velikovsky, 1955, p. 55, pp. 96-9, pp. 104ff).

Repeated impacts (material and electrical) would disturb the Earth in its orbit within the magnetic tube. The globe would wobble, the magnetic axis would constantly seek realignment, only to be subjected to another disruption as another megalith fell (Dachille, 1963) or a gigantic thunderbolt struck. The assault would crack the crust in many places (Norman *et al.*, p691), cause local uplifting, and alter the electric current in the outermost region of the Earth's conductive core.

Meteoritic fallout would range from microscopic nodules, similar to those found in the seabeds of later eras, to colossal intrusions of rock and/or metal. The Sudbury irruptive in Canada is an example. It is an elliptical ring sixty kilometers by twenty-seven, enclosing an asymmetric basin up to three kilometers thick. Along its boundary are large quantities of broken native and irruptive rock. This intrusion is judged to be younger than the rock surrounding it (Douglas, 1970).

The existence of ore mountains (isolated metallic deposits of mountainous size) like Marampa in Sierra Leone is also evidence of celestial fallout (Bellamy, 1951, p. 196). But the Sudbury basin and Mount Marampa are far from being the only examples of celestial intrusion: these are found on every continent, and certainly more astroblemes will be discovered.

Whereas the larger irruptives devastated local features upon which they fell, smaller pieces merely bombarded the surface without exploding, like artillery duds. People can survive intensive explosive barrages, as did most defending soldiers and civilians on Iwo Jima and at the Abbey of Monte Casino during World War II; pre-historic populations were no less survival-prone. Much of the smaller debris simply dented the surface and lay there exposed as testimony of a perplexing celestial activity.

When a material impact occurs, electric fields are produced, causing electric charges to flow (generating an intense magnetic field). Dachille (1979) asks:

> What mechanisms account for the changes in crater forms from the simple bowl to the awesome mare?

And then he replies:

> It should be noted that the microcraters observed on crystal faces or glass beads in lunar samples do not differ significantly from the Arizona crater or its lunar equivalents; the impact energies involved span at least *twenty* orders of magnitude. However, on progressing from bowl through the terraced-, peaked-, and melted-floor craters to the maria, the total energy difference amounts *only to six* [more] orders of magnitude.[79] This marked change in behavior can be related quantitatively to the reaction of the EM fields with the magnetic and dielectric properties of the target as a function of the duration of the EM pulse and the passage of the much slower shock wave pulse; in the upper range of energies the EM processes overwhelm the mechanical ones and thus determine the physical, chemical and petrological character of the resultant craters.

Spotting the Earth's surface are tektite fields. The large Australasian tektite field covers over five million square kilometers. From this field over 20 000 specimens have been examined.

Tektites are glassy spheres, of refractory materials, erosion due to air-friction melting as they fell through the atmosphere having depleted them of their less-durable components.[80] Tektites have rained down upon the Earth episodically since late Mesozoic times (presumably the Cretaceous), according to Baker (1960, p. 293).

---

[79] Bracketed word is ours.

[80] The tektites seem to have encountered the atmosphere (with present properties) moving at ten kilometers per second along shallow trajectories (Faul).

Chemical studies show that tektites resemble both terrestrial sediment and lunar soil, but significant differences distinguish them from both. To explain their deficiency in volatile material, the tektites must fall to Earth at velocities sufficient for friction-induced melting and scouring to cause chemical changes to their incipiently silicic composition; heating to 1475 K would produce sufficient such ablation (O'Keefe, 1978). Accordingly, O'Keefe has conjectured that the tektites were fired at the Earth by a hydrogen-powered lunar volcano. Equally, they could be products of the electric arc, or ejecta from the breakup of Super Uranus; more likely they were generated in cosmic thunderbolt strikes to Earth which occurred at intervals while Solaria Binaria disintegrated.

Tektites have been unearthed along with the fossil bones of Java man. Likely their falls were witnessed by prehistoric and ancient man and the spheres treasured as sacred. The experience would be remembered. In China, they were known as "fire-pearls"; and it is a "fire-pearl" that is pursued, in traditional representations, by the dragon, associated by Cardona (1976, pp. 42ff) with the memories of comets, possibly proto-Venus of circa 3450 BP.

Most of the meteoritic debris encountered by the Earth today is in the form of microscopic dust. Estimates vary a millionfold, but the Earth sweeps up a minimum of one tonne of dust per day (Singer, 1967). Daily falls of 44 times this amount are considered to be realistic (Hughes, 1976). A nine-year annual average gave 1.04×10^{11} grams (285 tons daily) in New Mexico sampling (Crozier, 603).[81] In two years the annual fall averaged 685 tons daily. Depending upon the influx and upon the timescale, the amount of meteoritic sediment can be calculated. Some scientists consider that a considerable fraction of earthy sediments (what amounts to about 3×10^{18} tons) are estimated to be meteoritic in origin (Niemann).

Most of this extraterrestrial dust must have fallen during outbursts in Solaria; at the present rate of influx, even allowing hundreds of millions of years since the Cretaceous, only one-millionth of the required meteoritic dust would drop: hence the estimate gap above. We conjecture, to conclude this set of guesses, that the Earth, from its primordial seed, could accrete from the plenum its present volume, less its sediments, in a millennium; its sediments could have been laid down in some generations of late binary times by extraterrestrial and turbulent surface events.

The observation in the infrared that some nova outbursts produce a significant silicate dust shell (Ney) leads us to suspect that the eruption of

---

[81] Spherules used in the counting measured 5 to 60 micrometers in diameter.

Super Uranus deluged the Earth with "meteoritic till", vast intrusions of dusty débris. In a short outburst the débris, which in some geologists' minds must have taken millions of years to sift down, might be plunked down upon the lithosphere. Donnelly (1883/1970) argues that vast fields of till scattered over the world are cometary fallout and not the remains of ice ages. It is more likely that both ice and till were of superterrestrial origin.

The first pre-nova eruption of Super Uranus probably rained down megaliths, rocks, glass, gravel and sand, but ice and water also fell from the sky in great amounts. The Earth was inundated with water condensed electrically from the plenum. Typhoons formed in explosions and towered into the plenum (de Grazia, 1981). They might occasionally be seen - roaring, stumbling pillars of smoke, water, electrical discharges and debris: veritable automotive disasters. New winds blew the waters across the face of the land. Since there is no compelling reason to suppose that great basins existed on Earth such as collect today's oceans, the flooding was severe. Some of the water drained into the craters blasted by meteorites and by electric bolts. Other waters slipped into the numerous fractures that appeared and into ponds fashioned by local thrusts and folds of the sediments.

An annual rainfall of two-and-a-half meters (not uncommon in coastal areas today) would dump over one million cubic kilometers of water onto the Earth's surface. This amounts to about 10^{18} tons of water, or about 1/3850 of the present oceans. Cherrapunji (India) receives 11 meters of rain in 159 days, which extrapolated (at the mean daily intensity) would yield 26 meters of rainfall annually. Hurricanes deposit rainfall at over seven times the rate at Cherrapunji; globally, such hurricanes could fill the ocean basins in five decades!

By current standards, a Deluge would constitute a more extensive rainfall than this. But for a biosphere used to Pangean conditions, where rain had been supplied by mists, the new kind of heavy rainfall would be traumatic.

The blast of material moving down the magnetic tube from Super Uranus created shock waves in the plenum. Where rarefaction occurred, water vapor froze, producing ice. Some of this ice fell onto the Earth. Within a short time ice sheets formed and grew all over the globe. Those were not polar ice caps. The ice caves of the intermountain plateau of the Pacific North-west region comprise millions of tonnes of ice (Patten), sandwiched between layers of lava. They are a surviving example of ice which fell from the sky. Clumps of ice avoided the numerous hot spots and lower altitudes of the world. As the ice continued to fall, electrical processes funneled most of it towards the magnetic poles, where large ice caps accumulated - this was the first ice to accumulate in what today we consider high latitudes. These polar caps grew and joined onto the sporadic patches, spreading rapidly towards the magnetic equator. The ice would probably have covered the globe and exterminated the biosphere had Super Uranus not erupted again.

CHAPTER TWELVE

QUANTAVOLUTION OF THE BIOSHERE: HOMO SAPIENS

Subjected to the effects of an unstable star, Earth's biosphere quantavoluted by extinction and genetic realization into the present form. To be emphasized here are the recent wave of genetic realization and the advent of *Homo sapiens* as an observer of the history of Solaria Binaria in its last stage.

Radiometric chronology and geochronometry based upon gradual stratification are incongruent with the model of Solaria Binaria. The fossil record, which is the guarantor of traditional geochronometry for the phanerozoic era, is generally acknowledged to be fragmentary, disjointed, and anomalistic (Ager, ch. 3). It is beyond the scope of this book to attempt a reorganization in detail of the geological and palaeontological record, and we have had to content ourselves with using conventional labels in a preliminary sketch of the route which such a reorganization would take. Table 6 exhibits in its first part what we would regard as the several significant major divisions of binarian history, leading into a more refined division, also contained therein, of the final very recent quantavolutionary times.

The Carboniferous appears in our view to have been a brief and thoroughly catastrophic set of episodes that bulldozed, burned, blasted, and buried masses of marsh and shallow water life forms in certain places, giving the illusion today that the whole (small) world of that day was a swamp. It should properly be assigned to the period of Super Uranus instability, a period of great extinction, rather than to a 65 million-year period preceding the Permian Period, where, significantly, boundaries are admitted to be rare.

Table 4
AGES OF SOLARIA BINARIA

Suggested Names of periods	Years Before Present*	Duration in Present Solar years	Description of period
A. Super Solaria	? to ~ 1,000,000	-	Electric cavity... galactic region depleted of electrons... space-material compressed into star... star transacts launching ion wind into space thereby increasing its electron density.
B. Radiant Genesis	1,000,000 to 750,000	250,000	Star erupts into binary at unstable epoch... strong inter-component electrical transaction... electric flow catalyses cell production... self replicating mitosis... biologic diversification of species and habitat.
C. Pangean Stability	750,000 to 14,000	736,000	Binary components separating... arc operating... biosphere thrives in plenum and planetary environments... biological creativeness declines.
D. Late Quantavolution	14,000 to 1,600	12,400	Arcintermittent... plenum thins... binary becomes unstable... planets isolated, devastated and relocated as binary translates into Solar System.
I. Urania	14,000 to 11,500	2,500	Deluges form icecaps and floods... breakup of sky canopies... Homo sapiens schizo-typicus appears... ecumenical culture... Uranus Heaven religion.

II. Lunaria	11,500 to 8,000	3,500	Global explosion and cleavage... Moon eruption ... ocean basins formed and filled... displaced continents... biosphere quasiextermination... people isolated and fully traumatized... lunar worship.
III. Saturnia	8,000 to 5,700	2,300	Biosphere multiplies... cloudy atmosphere... no ice caps... settled continents... expansion of regional cultures.. rich technology... Saturn worship.
IV. Jovea	5,700 to 4,400	1,300	Noachian shelf floods and high tides... lightning and cleared skies... new icecaps form... more severe seasons... dryclimates... eastward movements from "Atlantis" to Egypt and Mediterranean ... empires form amidst widespread conflict... Jupiter worship.
V. Mercuria	4,400 to 3,450	950	Separation of magnetic and geographic poles... axial tilt enhanced... pyramid age... large new civilizations in Mediterranean, China and Caribbean... Olympian family worship.
VI. Venusia	3,450 to 2,775	675	Devastation of globe by protoplanet Venus... religions and cultures reduced and remodelled... Venus worship... large petroleum fall-out.
VII. Martia	2,775 to 1,600	1,175	Mars Earth Moon and Venus transact destructively... war-like

				cultures promoted... Toltecs, Myceneans and Etruscans reduced... Mars worship.
E. Solaria	1,600 to 0**	1,600		Settling of present Solar System... secularization, philosophy and empirical sciences ... synthetic religions.

*2000 AD = O BP.

** Solaria is defined to begin with victory of Christianity in the Roman World, eclipse of the pagan gods and their appropriation by solar imagery.

Most of the earlier Silurian, Devonian and Permian periods would fall into our middle category of Solaria Binaria stability.

Even earlier periods of the controversial scale are assigned to our period of radiant genesis. The scarcity of fossils in early Cambrian rocks indicates their formation and turbulent experiences in the early radiant period.

Originally, geologists and paleontologists hoped to trace natural history backwards through the rocks and establish a long chain of rock-related fossils on the principle of super-position, the first and perhaps only quite defensible concept of natural history. Such hopes were dashed early, but the ideological stimulus behind them was so strong as to obscure the fairly obvious origins of rock and fossil discontinuities.

Discontinuities (*unconformities* is generally synonymous) imply quantavolutions, whether treating of rocks or fossils. No continuous column of rocks or fossils exists. All => *fossil assemblages* that incorporate flora and fauna of diverse life niches, as a flying animal and a fish, or a hippopotamus and a reindeer, are evidence of quantavolution. Logically, and for other reasons, the rocks that contain them have been quantavoluted at the same time. Traditional geochronometry, already in a crisis of self-doubt, compromised with the new science of radio-chronometry, allowing itself within this century to move from a forty million-year to a 4.5 aeon Earth history. This thousand-fold increase was accepted on the assurance that radio-isotope fractions can be used as a clock, if the initial balance of the isotopes is known. Such is not the case, as even the eruption of Mt. St. Helens showed in 1980 (Rawls). Besides the trenchant negative criticism of radio-chronometry (Cook, 1966, pp. 23ff), the modes of genesis and agglomeration of the Earth invoked in the present study supplant the kinds of elemental mixes presumed by nebular models of Earth genesis.

Recently more direct attention has been accorded the waves of extinction that typify the fossil record (Valentine, Raup), and the theory of extraterrestrial causes of extinction has entered the house of science from its stable as a *Grenzwissenschaft* (fringe science). Massive intrusions of solar protons have been postulated as the cause of the extinctions and accompanying mutations (Reid *et al.*, p179). In the period after the Mesozoic, the collision of cosmic bodies with the Earth has been proposed as an alternative explanation for the extinctions (Urey; Alvarez. *et al.*).[82]

Known living species number upwards of one million; estimates of living but unidentified species may reach to eight and one-half million (Passerini). The number of different species since the beginning of life was estimated at five hundred million by Simpson (1952). Fossilizable species were estimated at ten million by Teichert, of which nearly half would be marine (Passerini), but only some one hundred and twenty thousand fossil types have been identified. Thus, one in fifty species would be fossilizable, and one in a hundred of these, or 1 in 5 000 of all pre-existing species, would now be known. It may be argued nevertheless, as has Cook (1966), that the fossil record is relatively complete, and that the fossils already discovered form the vast majority of pre-existing species.

Clearly, the definition of species, both as to those living and those extinct, must greatly affect the numbers. Further, in biological development speciation is much less important than major changes, as indicated in definitions of phyla, classes, orders, and families, but especially in definition of the stages of development of the living cell. Major natural change has probably ceased. Much speciation will probably come under human control, even as existing species will continue on their course of extinction. The history of Solaria Binaria would not promise the species a reprieve; this, if granted at all, must come from the laboratory. Humans are a part of the problem, being themselves in a posture of self-extinction; hence, the laboratory work may begin with the laboratory workers.

The biosphere, when Solaria Binaria began to degenerate into the Solar System, was at a stage roughly equivalent to that which has been denominated in paleontology as the Triassic. All major life forms of today and most of their families and species were identifiable, but many species were absent, including the human. Conventional reckoning has already

---

[82] Such collisions would, as we have shown, cause magnetically confused sediments to be laid down, at the times of bombardment. Sudden biological extinction has been linked to periods of magnetic confusion in the paleological record (Whyte).

moved *Homo sapiens*, defined as an ancestral hominid working with tools and building shelters, back by between five and ten million years into the Cenozoic. Under such circumstances, he would encounter extinct reptiles, mammals, fish and birds, and travel between continents over broad land bridges now inundated. It is not expected that the human age will ever reach back to the Triassic, but it may be that the Triassic will reach up to the human. This may happen by assuming - with whatever adjustments may be required in the interpretation of the sporadic fossil record - that almost all present families and species, if not existent prior to the Period of Quantavolution, realized themselves in this period; most at the beginning of it, 14 000 to 10 000 years ago, some even later. It may be that the now well identified Permo-Triassic extinction was the period of Super Uranian novas (14000 to 10000 BP).

Figure 26
Radioactivity of Fossilized Remains.

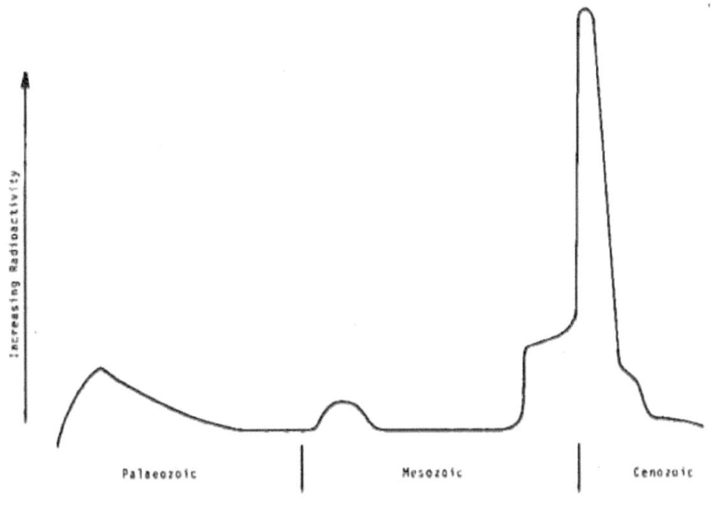

Evidence from several widely separated investigators indicates that fossil remains from the Upper Cretaceous are highly radioactive. Reptile bones containing as much as 0.11% U_3O_8 have been found in Brazil. Fossils ascribed to earlier eras show much less radioactive content than remains dated at the Cretaceous - Tertiary boundary.- Figure after Kloosterman

At this biological discontinuity Raup calculates a loss of 13.5% of the classes, 16.8% of the orders, and 52.0% of the families of well-skeletonized marine vertebrates and invertebrate animals, and of 64.8% of the invertebrate genera. He reasons that 96.0% of the species of echinoids were extinguished then, too. Basing his estimate upon a standing species diversity of between 45 000 and 240 000 in the Permian, he concludes that the marine biosphere would have been left with between 1 800 and 9 600 species, from which the present species come. We call to mind that earlier we proposed a desiccating climate for the epoch when the plenum declines; the extinctions noted may be related to this phenomenon. The later extraterrestrial discharges of water collected into deep pools rather than in shallow marshes, once the ocean basins were sculpted. The end of the Triassic sees further mass extinctions. So does the cretaceous, which concludes with the disappearance of the dinosaurs and other groups.

In the Cenozoic, "speciation was rampant, as a multitude of niches was invaded in the replacement of extinct reptiles" (Stanley). An average species of late Cenozoic mammal survived one to two million years without transitional forms. With this average, it seems impossible to account for changes from primitive forms to bats and whales, in twelve million years of the early Cenozoic. So reports the same author, who notes that "much more than fifty percent of evolution occurs through sudden events in which => *polymorphs* and species are proliferated". In the American West, drawings of dinosaurs are said to have been found, presumably by the hand of ancient Indians (Hubbard). The existence, on the banks of the Puluxy River in Texas, of human footprints (not detectably different from the footprints of a modern human) in sandstone alongside dinosaur tracks would make the coexistence of humans and dinosaurs hard to dispute. At the end of the Pleistocene, conventionally tied to the last Ice Age and dated ten to fifteen thousand years ago, another wave of extinctions struck the biosphere (Martin and Wright).

In view of these mass extinctions, any lingering hope that an evolutionary record can be completely displayed and then proven must be abandoned. So must the similar hope of proving an evolution of the lithosphere using fossils. No continuous stratification either of fossils or of rocks exists. Under these conditions, where discontinuities and unconformities mark the geological fossil record (Ager, ch. 4), quantavolution becomes the ruling concept. Fossil and rock discontinuities are to geological age boundaries what ruined settlements are to Bronze Age boundaries. Originally established to show transitions or anomalistic happenings, they end up as benchmarks of disasters. Further, the omnipresence of fossil assemblages as the basis for paleontological studies of succession is a sword of Damocles over the head of evolutionist. Fossils, themselves, are creatures of personal or, usually, of collective catastrophe.

No new life forms are attributable to the interval of the Pleistocene extinctions. It may be that few new forms are associated with any extinction of the third and last period of Solaria Binaria.

Apart from ideological hopes, two processes may have served to give the impression of new species and families evolving at or between extinction events. One is the bias of the fossil record, which rewards large numbers and calcium-bearing superstructures with a badge of existence. We believe, rather, that almost all modern species have survived from the Period of Radiant Genesis, either in their present form or in a form carrying in its germ plasm the present form and intervening forms awaiting realization.[83] Under catastrophic conditions immediate mutation and adaptation are possible among some individuals. Thus in a sense they both perpetuate and generate a species. Hence, non-populous species can have persisted all along and appeared in the record when their populations expanded under the "right" conditions. Further, these species and other species already part of the old (Devonian) record have quantavoluted into "new" species under the same catastrophic, mutative, and adaptive circumstances.

The difference between the certainly catastrophic age of radiant genesis and the catastrophic recent record of explosive quantavolution clearly rests in the extremely powerful and rich environment of the first period and its vast domain of the plenum. The period of collapse of Solaria Binaria was incomparably poorer in genetic capabilities; to extinguish, yes; to capacitate, also yes; to create, no. The many millions of mutations and environmental changes occasioned by the instability and destruction of the system were paltry by comparison with the possibilities of the first period.

Therefore, when one approaches the subject of the genesis of Homo sapiens, one need not expect grand changes of a bio-physiological type; these do not exist. With protein "chains" as the basis of comparison, humans and chimpanzees "share more than 99% of their genetic material" (Washburn, p203). A comparison of the earliest fossils of hominids with the similar parts of modern humans does not demand an acknowledgment that the two are of distinct species; and judging from remains alone, the hominid may have equal or greater capabilities than the modern human. For example, the brain case of hominids, which may contain 500 cc. is not so small theoretically as to preclude intellectual competition with a modern human brain. Though larger by far on the average, modern mankind does

---

[83] This may be recognized as related to the concepts of "paedomorphosis" and "clandestine evolution" (see *Ency. Brit.*, 1974, Macro, 19).

offer braincases that, while intellectually competent, are akin to the hominid's in relative size. This is quite apart from the presently unresolvable issues of the intensity of convolution of the brain and the percent age of brain tissue ordinarily utilized.[84]

The view here conforms to the theory of genetic realization. It may be maintained that hominid is as old as the end of the period of radiant genesis; further it may be maintained that hominid had a genetic potential for becoming the modern human. A large change is not necessary to differentiate the human from the hominid.

It would appear futile to search for differences in traits that recently socio-biologists have already discovered in other primates or animals: sociability; group obligations; signaling; using sticks, building houses and nests; organizing expeditions; intricate social bonds; and so forth.

It may be equally futile to seek after biological differences; manual dexterity; bipedalism; brain size; omnivorous dentition; and so on.

Perhaps the most rewarding area of research would be in the mechanisms that govern traits most peculiar to humans (although least likely to be determinable from fossil remains). Most peculiar to *Homo sapiens* from his earliest appearance has been a "non-trait", his generally defective instinctive structure. Active fear and self-awareness resulting from it generated his symbolic and ideological behavior. These are logically connected, as has been shown in detail elsewhere (de Grazia, 1983b, 1983c). Their mention here helps to explain how it happened that we have human testimony to use in constructing a natural history of Solaria Binaria and the extent to which such testimony may be reliable and valid. The simplest change would be a general constraint upon instinct. Instinct is a non-learned activity and response, unfettered by self-awareness. *Homo sapiens* is the least instinctive of all animals, hence the least predictive and most responsive to internalized planning. Very many, perhaps all, human actions and physiological processes can be internally constrained or modified unconsciously (psychosomatism) or consciously. The extraordinary achievements of *Homo sapiens*, it is argued, are entirely due to the operations of an instinctual incapacity upon an otherwise normal primate constitution. This instinctual incapacity is closely connected with and may have given rise to the generalized anxiety or fear characteristic of humans, especially "intelligent" humans. Human fear, resting on top of animal fear, was originally fear of oneself, fear of the inability to act and react instinctively under conditions of the mental division of the self into

---

[84] Modern humans can function broadly and intelligently on half a cerebrum, one hemisphere.

several differently aware parts.

The transformation of hominid to human with respect to instinct delay, which leads to self-awareness, which then promptly adduces symbolism, ideology and recall, is most likely to have been accomplished by contradictory pressures - one to diminish instinctive response and the other to increase response. Together they produce continuing anxiety and a number of mechanisms to cope with it.

Some of the pressure to diminish instinctive response may be attributed to an increase in electrical resistance between the two hemispheres of the brain, distributed throughout the *corpus callosum*, the large membrane occurring between the two hemispheres. This membrane would increase its resistance to the passage of messages between the right and the left brains, which are in fact electrified and responsive to changes in the external and internal environments.

An environmental de-electrification would seem to occur as the Earth's interior increased its supply of electrons (relative to its cosmic surroundings) simply by the steady accumulation of charge. In a changed environment, the repetitive correlating signals that constitute a large part of the exchange between the two hemispheres of the brain would encounter increased environmentally induced resistance; so they would bunch up and interfere with one another. That is, fewer transmission lines would be available to the same number of messages.

The brain originated in a world of lower electrical levels and greater electrical differences. There may be a functional problem today in a world where electrical levels are higher and electrical differences much diminished.[85] The brain was possibly originally more stable, that is, instinctive, perpetuating the less anxious hominid.

The messages between the brain hemispheres propagate relatively slowly, by direct current through chemo-electrical diffusion, so to reflect a slightly diminished electrical constant, enough to furthermore "encourage" crowding of signals and a more frequent de-synchronization. The effect would be both delay and confusion - delay in microseconds in assessing a neural trigger for an information or command bit, and confusion in overburdening the channels with combined but incompletely co-ordinated messages.

Signals that must "wait" and may get out of phase would necessitate momentary verification of otherwise instinctual responses, a delayed

---

[85] This may account for some of the three-fold growth of the brain by comparison with fossil hominid.

reaction, and even conflict and aborted decisions. This is enough to set up the unique pattern of human behavior in an otherwise pedestrian mammal.

Thereupon two paramount qualities of the human mind would result; the need to think before acting, and the analogizing of experiences and events, leading to synthetic combinations of all types. In addition, we admit the possibility of a change in the functioning of hormonal glands, such as the adrenal cortex. A continuously higher level of secretion and induced stress - a new constant - might have been provoked by the disasters of the time of humanization and/or by a new, stronger and persisting electrochemical stimulation. The brain would be permanently stressed towards anxiety and action. Taken with the decline in the correlation of the hemispheres, this contradictory stress would further humanize the person with the evermore-poignant auto-instructions to "look before you leap" and that "he who hesitates is lost."

Promptly there would emerge a conception of the self, a continuous fear of loss of self-control developing out of the need to compromise with oneself, an aggression against those who provoke difficult decisions or restrictions of the self-conflict or who "cause one to have to think," and the need to talk to oneself (one's other self), which leads quickly to talking to others to engage them into talking to one's self - which leads in turn to talking to "the most important people in the world:" the anthropomorphized gods. The self would project its hopes and fears to the external world, but especially and exactly to those features of the external world from which the most impressive experiences emanate - the heavens.

Thereupon the human mind is structured and in place. The devising of culture was practically instant. Words, operations and thoughts establish social contact on a level unknown to "hominids", and a "social contract" comes into being. Society helps people to talk to themselves; people talk to themselves through other people.

The social process, the instant culture, is not only formed of the present. It accrues memories. It recalls. It is obsessed with its own creation simply because it is so unbelievable and dramatic (traumatic). Since this scenario was enacted only 260 => *memorial generations* ago (de Grazia, 1981), the transmission of some valid and reliable information in decipherable form need not be surprising.

Humanization and culture seem to have appeared in the initial phases of Solaria Binaria's collapse, around thirteen thousand years ago, allowing for a thousand years of environmental instability to finally "get through" to the hominid, as described. The fact that all races share the human mentality indicates that they share a single ancestral line; no one has discovered a feral tribe or a live hominid. Still, because of the quasi-environmental character of the "mutation", several lines might have originated independently from individuals or groups hoarding the genetic substructure of the newly expressed trait. Whatever the case, the fact that many, perhaps all, peoples possessed an ecumenical "creation culture" would point to a worldwide takeover by a single culture within a thousand years.

The earliest human stories reveal something both of the character of the storyteller and of the events about which he speaks. Creation legends (and many creation legends remain unclassified as such) recall a time far before the time of their recounting. As a consequence of the need to control himself and his environment, the human promptly invented history, that is, a purposive and selective recollection of all that had happened to his group since he stood as a human upon the Earth (Eliade, 1954; de Grazia, 1981,1983b). Invariably the history began with a celestial disruption of an even-tenored, hardly conscious existence, or with gods preparing to destroy the primeval world in order to reconstruct a new world suited to mankind.

The catastrophic natural frame in which the hominid quantavoluted matched the terror that seized him as he humanized. It is the oxymoronic quality of this fact that has led most experts to question the ability of a catastrophized mind to report anything but catastrophes; they view catastrophic reports of natural history as the fictions of a savage mind - a catastrophized mind (which it is, rather than savage) prone to elevating personal problems into gross slanders of calmly evolving nature.

This position cannot be maintained in the context of the massive sublimation exemplified in legend, myth, fables and rites. If primeval man were "spinning yarns" in contradiction or exaggeration of actual happenings, he would probably tell stories with peaceful plots and happy endings. He would not incorporate gods, or even believe in them. Instead, he builds the totality of his culture on a tragic plane; sacrifice, suffering and punishment are its principal themes. The leading actors in his tragedy on these themes are always gods of the heavens.

Of all four possibilities, then, that refer to the experience of primeval man - catastrophized mind transacting with calmly evolving nature; calm mind transacting with calm nature; calm mind transacting with catastrophic nature; and catastrophized mind transacting with catastrophic nature - it is this last that appears to be closest to the truth. Catastrophically originated, *Homo sapiens* built upon his irrepressibly fearful and scarcely controllable mind. With this mind, he observed and recalled with obsessed determination the time of his creation, and all subsequent landmarks of history that reminded him of the circumstances of his creation.

CHAPTER THIRTEEN

NOVA OF SUPER URANUS AND EJECTION OF THE MOON

Ancient Mesopotamian accounts of gods tearing off each other's heads and limbs are not "baffling" (de Santillana and von Dechend, p. 303) in the context of early human existence. But these and similar stories in the Teutonic, Greek, Roman, Hindu, Iranian, Mexican, Egyptian, and archaic (" primitive") religions are baffling in regard to their positioning in time. Given an empirically established calendar, a general review of the early literature may assign a period to them. Tentatively, we assign these earliest theomachies to the period of Super Uranian instability and the climactic nova of Super Uranus that drastically changed the face of the Earth

According to Hindu accounts (Brown, pp. 281-9), while Adityas and Vitras fought in that troubled first phase of the skies, Heaven and Earth, living together in a common house, bore Indra. At first concealed, he fed upon soma until he attained enormous size, whereupon he blew Heaven and Earth apart forever, filling the atmosphere by himself and exploding the Vitras in the process by thunderbolts. From the exploded belly of Vitra came the cosmic waters, acknowledging Indra (Super Uranus) as their new lord. Out of the waters came also the Sun. Varuna (Heaven as Super Uranus) presided, as order and truth emerged from primordial chaos.

This narrative is but one culture's account of mankind's witnessing of the explosion of a celestial body. An alternative, from the Vedic period, has a Cosmic Egg (here Super Uranus) floating for a thousand years in the primordial waters (our plenum) until it burst (as a nova) to reveal the Lord of the Universe, Purusha. It may be that Purusha is yet another phase of the troubled Super Uranus.

From the navel of Purusha sprang a lotus bright as a thousand suns (possibly the electric arc), whence came Brahma, who acquired Purusha's powers as Lord of the Universe (and whom we shall identify below as Super Saturn) (Cardona, 1978a, p. 43). The cracking of the Cosmic Egg may represent the sight of a fissioning Super Uranus in human memory.

The Hebrew Book of Genesis begins with a primordial light that did not have the company of the celestial bodies until "the fourth day". This may have been "lightness" or "a light". The deity may have been Super Uranus, who first gave "lightness" and then "a light" of himself. Some (e. g. Cardona, ibid.) place the deity here already into the Saturnian period, justifiably asserting the parallel names and qualities of Elohim and Saturn. We speculate that either Genesis begins after the Super Uranian nova, with Super Saturn, or that in the great expanse of time, Elohim in his Uranian role was merged into Elohim in his Saturnian role. Generalizing on this problem, Tresman and O'Gheoghan comment that "where there is descent (from father to son) it is obvious, otherwise the transition between the original deity and the later Saturn god is not too marked."

"The Sumerian ideogram for 'star', 'god', and 'heaven' (An) is one and the same, a simple eight-pointed star shape. This strongly suggests that they believed that the original "heaven" was a body that later became a star. It also strongly indicates that the first deity was this star-heaven god. (Tresman and O'Gheoghan, quoting Kramer, 1963 and Peter James, p. 36) The Egyptians defined Atum as "the incomplete one who became complete," says Lowery (*ibid.* fn., citing the Coffin Texts); we may surmise this as Heaven becoming a star; Atum was depicted by the Egyptians as a setting sun (Ions, p. 40).

"There is every indication that this original deity was at one time the only visible planetary body of the heavens. From the Hindu sources we have: 'In the beginning Prajaparti existed alone'. The Egyptian records tell that Atum 'was alone in the primeval watery abyss'. The deities An/Anu (Sumeric) and Ouranos (Greek) were both lone planetary deities, although their names translate literally as 'heaven'. In each case the successor to the original deity was a Saturn-type god." (Tresman and O'Gheoghan, p. 36)

In the *Edda* epic of Scandinavia, "The Spirit brooding over the dark, abysmal water calls order out of chaos, and once having given the impulse to all creation, the First Cause returns and remains for evermore *in statu abscondito!*" (Blavatsky, citing Mallet and the *Edda* epic, pp. 160ff)

The star, in its pre-nova state, was apparent from the time of its emergence from out of the gloom into the now activated heavens. The interval from 14 000 to 11 000 years ago may be designated as the Age of Urania. The skies were falling upon the transformed primate schizoid, *Homo sapiens*. Children's fables like that of "Chicken Little," who led the barnyard animals in a search for an Authority to do something about the falling skies, are ancient and widespread and are not to be neglected as reflections of the ancient traumas imprinted upon the collective memory and sublimated into the first fictional literature alongside the sacred religious myths (de Grazia, 1978, 1984a).

When the heavens were broken open, as by P'an Ku, the Chinese creator god, Super Uranus appeared in the north, immense and egg-shaped, probably resembling a giant eye, too, atop the sky. At first the white of albumen, it became yolk-red, and radiated heat. It was probably the primordial light of the beginning lines of the Hebrew Genesis; as noted above, the present celestial bodies appeared only on the fourth day of creation.

To the south, less luminous because it was much more distant, was the Sun. There were now three sources of heat, the magnetic tube (powered by the arc), the Sun, and most prominent of all, Super Uranus. Depending upon the earthly observer's location, either Super Uranus or the Sun could be discerned through the thinning gases. In general, northern observers glimpsed Super Uranus, southerners saw the dimmer but larger Sun.[86]

During its period of instability Super Uranus erupted regularly. The body of the star contained electric charges distributed internally to be in balance with the charge on the surface, which was transacting with the Sun and/or the Galaxy. The Sun via the arc had been robbing Super Uranus of electrons for almost one million years, a process that kept the surface of Super Uranus relatively drained of charge.

Super Uranus' charge-deficient surface could be altered in one of two ways: a sudden influx of Galactic charge, or a short-lived disruption of the arc. Both would "overcharge" the Uranian surface. The reaction would be to further concentrate the charges within Super Uranus. When the surface charge again became reduced (which would happen if the arc suddenly reconnected, allowing a burst of ions onto Super Uranus) the interior of Super Uranus, now overly packed with charge, would respond by outbursting charged matter into space.

This outburst would not normally escape from the domain of the star which generated it; most of it would in time be reabsorbed into the star, returning the "released" charge. Charged debris falling back could help induce conditions for another electric compression and outburst. And so the star erupted cyclically.

The cycle of erupting away highly charged material and subsequently

---

[86] At some latitudes both bodies were visible, either together or alternately. We believe ancient accounts of two great lights in the sky refer to this era, or to the later Saturnian era (before the Deluge).

re-absorbing it follows directly the notion of a plenum of charged gases around the binary system itself and secondarily around each charged body of the binary.

The extent of each plenum is determined by the charge on the body it surrounds and by the charge in the plenum gases. When an eruption occurs the plenum gases increase in charge and expand their sac. Electric fields are set up which cause charges to flow, thereby decreasing the surrounding charge relative to that within the central body. The plenum then begins to collapse, pieces within the sac discharge and also fall back. In the Age of Urania each body might be said to possess an autonomous sac and plenum, immersed in the now diluted plenum of the whole system.

The plenum of Solaria had by this time become so tenuous that the individual bodies had established around themselves electro-spheres - regions of charges, gases and, from time to time, solid matter. These spheres, or regional environments, were to their central bodies what the plenum had been to the system, in that they defined the limit of the body's influence upon nearby matter and charges. Vestiges of the electrospheres are found today in the electric sheaths surrounding the Sun and the various planets; the transition occurred mostly in the time of Jupiter (see ahead to Chapter Fifteen and Note B, fn. 117).

The resorption of erupted pieces occurs so long as they do not exceed a certain critical size. Judging by today's Solar System, this would seem to be about 22 kilometers in diameter or a volume of about 6 000 cubic kilometers. Pieces that could not be resorbed could become satellites of their parent body - as had Super Uranus and the primitive planets - but in Solaria they could be transferred from the realm of one body to another whenever the two electrical plena involved were contiguous. In these special circumstances escape was possible; the smaller sacs could leak gases and pieces into adjoining sacs. This is how Super Uranus bombarded the planets. Its outbursts electrically charged the Earth's sac and filled it with debris of diverse sizes.

The cooling of the Earth, noted during the Uranian Period, could be accomplished by several means. As the plenum cleared, more and more of the arc's energy was transmitted directly to the Earth without involving the plenum gases as intermediary. Thus => *albedo* became more important in the energy transfer. Where earlier the heated plenum kept the Earth warm, now the light conveyed energy directly to the Earth. If the cloudy Earth reflected 52% of the radiation from the arc the Earth would cool to 270 K from its former warmer temperature (see behind, Chapter Six). Alternatively, if the Earth accepted more than half of the light but the arc cooled, the temperature would also drop.[87]

In its climactic explosion, Super Uranus ejected a large chunk of its material down the magnetic tube towards the Sun. This element, to be termed Uranus Minor, was preceded and accompanied by gaseous blasts and water. Badly out of electrical equilibrium, both because of the electrical cataclysm which ravished Super Uranus and because the ejects now followed orbits taking them into regions of greatly different space-charge, a vast, brightly glowing space-charge sheath surrounded Uranus Minor as it hurtled towards the Earth (see Juergens, 1972, p12, for a discussion of these sheaths).

At the time of the eruption the Earth is revolving around the arc, moving counter-clockwise (viewed from Super Uranus). The globe is oriented with Africa (the old north) facing the explosion; the magnetic poles lie on the rotational equator, Greenland leading and Antarctica following.

At this time the Earth had a continental crust everywhere. The continents that survive today were bunched around Africa, then located at the north rotational pole. In Figure 27 the land areas of the world today are drawn schematically as they related in Pangean (all-Earth) times. They accord with the geophysical and paleontological findings of the continental-drift school of thought.[88] Uranus Minor, moving from Super Uranus towards the Sun, encounters the outside edge of the Earth (the Pacific side) in passing (see Figure 28).

An intense transaction occurs between the two. Electrical polarization distorts the shape of both bodies and their sacs, and the Earth's magnetic axis wrenches out of line, which causes the world to shudder. The sudden movement loosens part of the lithosphere; torrents of water (or ice) flow (or slide) across the surface. Fiery blasts strike the area which is now the West Central Pacific Ocean, opening massive craters, some deep enough to release mantle material previously thirty kilometers below the surface.

At perigee the transaction between Earth and Uranus Minor reaches a maximum. The crust on the side closest to the intruder cracks and fragments. Explosively, as much as half of the Earth's continental material rises into the sky, leaving exposed much of the upper mantle. This

---

[87] An Earth reflecting 30% of the light and an arc reduced by 32% would also cool the Earth to an ice-age condition. The actual mechanism by which the Earth receives its => *insolation* is open to question. Hanson notes that measurements made from space have necessitated lowering the Earth's global albedo from 45% to 29%, => *irradiance* values to the surface having been raised by up to 27%.

[88] It will become apparent that our theory (cf. de Grazia, 1981, 1984b) posits a continental rapid rafting of a thousand years or so, rather than the usual 200 My drift.

extraction initiates the reshaping of the lithosphere to produce the structure we study today. Whereas before this event the entire Earth was topped by a thick granite layer, basalt was now exposed. In its brief encounter with the Earth, which we estimate to have lasted many hours, Uranus Minor peeled a deep swath of crust (and some upper mantle material) from the Central Pacific and to lesser depth from the great seamount area west of the Americas.

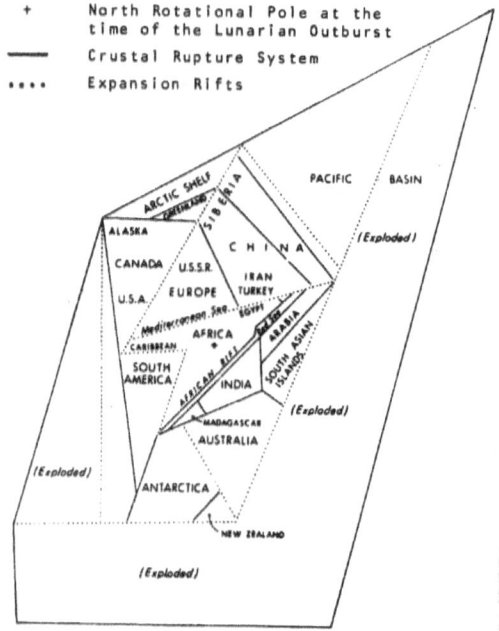

Figure 27
The Surviving Land from the Age of Urania

Prior to the eruption of Super Uranus, which hurled the large fragment Uranus Minor down the magnetic tube past the Earth, our planet was covered by a complete shell of granitic crust. Much of that crustal layer was lost when the Earth encountered Uranus Minor. That crust which remains was once clustered around the ancient North rotational pole, which then always faced towards Super Uranus. It was ruptured and rifted by the close passage of Uranus Minor.

The Earth wobbled eccentrically as mechanical, electric and magnetic forces acted upon it. Surrounding the wounded surface was a rampart of devastated granites. Within it were thousands of seamounts, unable quite to explode into the sky, and now frozen like pulled taffy. At its center was

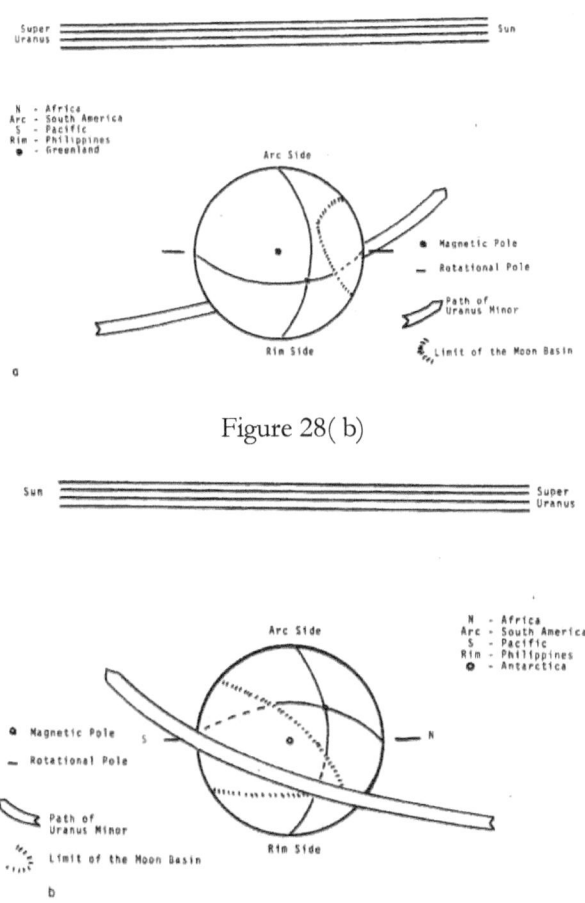

Figure 28(a)
The Encounter of Uranus Minor with the Earth

Figure 28(b)

The Earth was in orbit around the electric arc at the time when Uranus Minor passed close by it. Then the Earth's surface was irrotational with respect to the arc so that one point on the crust (magnetic south) was always leading on the orbit (see above) and the point on the opposite side (magnetic north) was always trailing (see over page). Though the crustal arrangement of that time placed the lands differently, the lands on the present globe, which did not take up their present positions until after the catastrophe described here, would be oriented as indicated on figures 28(a) and (b) relative to the arc and the Sun Uranus Minor met the Earth on the trailing side and departed on the leading side (compare with Figure 18), tearing away crust and creating the Moon Basin where shown.

an abyssal plain where the surface of the Earth's mantle appeared scoured of its covering; it was then, and is now, the deepest basin of the Earth's surface. The blow-off was so great that it pulled the great central magnet of the Earth 436 kilometers towards it, making the shortest circle line of the Earth's magnetic field of today pass through the Society Islands. The Earth's center of gravity was pitched five kilometers towards the great Pacific depression (Baker, 1954, p. 5).

In this one brief event, the entire original Southern Hemisphere of the Earth's crust, was electrically ejected into the sky. The unbonding of the crustal granite and mantle from the subsurface magma involved a large transfer of energy. If the inter-body transaction is translated into thermal terms, the heat would have been perhaps impossible for the Earth to support without vaporizing the biosphere and the globe itself. About 10^{32} joules are theoretically required to peel off the surface layer of the Earth entirely. Here, over half the crust was ejected, but the balance was loosened and set into motion.

However, the transaction consisted of a trade of material for electric charge. Uranus Minor, much more heavily charged, deposited charge upon the Earth. The new electrical energy was incorporated by the molecules of the Earth. Their internal (atomic) bonds were stretched. A very large amount of energy was required by the chemical bonds and supplied athermally[89] by a huge column or front of lightning bolts blasting a swath into the sky during the pass-by. That heating which occurred was concentrated at the interface of crust and mantle and at the bottom of the moon basin. In adjusting its figure following the ejection the remaining land mass fractured and the Earth expanded by about twenty per cent (de Grazia, 1981; see also Meservey, p. 611). This represented a radial expansion of nine per cent and a corresponding atomic expansion throughout much of the Earth.

The remainder of the continental mass that had covered the Earth fractured into the complex ocean-ridge and land-rift system viewed today (see Figure 29). The separated blocks were electrically repelled[90] and squeezed apart. They rafted speedily towards the Moon basin. Lava welled up from below the fissures and widened them. Thousands of new volcanoes were instantly activated.

---

[89] That such an athermal encounter is possible is attested by the survival of a tree at Lugano, Switzerland, which did not ignite upon being struck by lightning. By a serendipitous coincidence the whole event was photographed by a scientist conducting research on lightning (Orville).

[90] Each continent is currently located antipodally to an ocean (Harrison, 1966).

The constellation of fractures exhibited in the world map of Figure 29 probably occurred within a day's time (de Grazia, 1981; Manson, ch. 4). Tectonic plate theory today relegates the fractures to a remote unspecified era, with ocean basins always present. It invokes various mechanisms to accomplish over great stretches of time complex slow movements of a number of plates carrying continental crust. The theory is not only unnecessary; it is mistaken on the most obvious criteria. Melvin Cook (1966, p. 189), from the perspective of his research on explosives, points out readily the unified and simultaneous features of the global fracture system, finding in them what is ordinarily to be perceived in an explosive impact upon a globe. Possibly, certain minor fractures branched out or lengthened in the following months or years. Some fractures were not fully consummated, such as the African-Near East rift and the trans-Asian rift. Others have been covered in part by subsequent torques of the crust, as in the case of the San Andreas fault, which was buried in the "westward" movement of North America, or the Red Sea-Adriatic-Rhine Valley rift, which was partially overridden by the Alps. It will be noted, too, how the Atlantic Ocean crack probably shot out from an Arctic base, traveled swiftly but against resistance, and then branched off to circumnavigate the south Pacific area, sending four continents on their separate journeys: South America, Africa, Australia and Antarctica.

A deluge of water fell from Uranus Minor as it passed. These waters more than replaced the water carried away with the lost crust, fell into the hot fissures and onto the volcanoes blasted into existence at the passage. There the waters exploded, expediting further cracking of the continents. The world, which shortly before had been heading for an icy end, now became hot and steamy and was threatened by falling water and rock.

A cubic kilometer of Earth's atmosphere at present contains ten thousand tons of water. If the Uranian deluges were precipitated by electrical activity of the Earth's electrosphere (see ahead to Note B), only 540 tons of water per cubic kilometer would be required in order to achieve the oceanic levels that we estimate occurred in the Uranian Lunar periods. In the course of a millennium an annual rainfall of 1.2 meters depth would have descended upon the Earth's surface, a rainfall that a substantial section of the world's people enjoy today. This would be adequate to fill the basins up to the continental slopes, about half of the present ocean volume. Most of the rest of rainfall belongs to the story of Saturn (Chapter Fourteen).

Figure 29
The Fractured Surface of the Earth

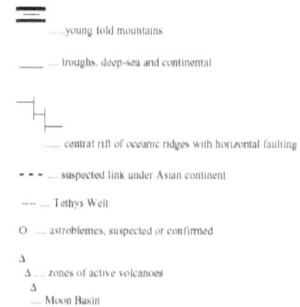

Key to the map:

— young fold mountains
— troughs, deep-sea and continental
— central rift of oceanic ridges with horizontal faulting
- - - suspected link under Asian continent
—·— Tethys Well
O astroblemes, suspected or confirmed
Δ zones of active volcanoes
Δ Moon Basin

Following the removal of the Moon from the bulk of the Earth by the action of the passing Uranus Minor, the surviving broken continental crust of the Earth shattered and rapidly scattered, taking up new positions around the crater surrounding the Moon Basin in what is today the Pacific Ocean deep. The troughs, ridges and faults of the current crust were sculptured as the Moon was torn from the Earth and as the Earth recoiled and recovered from that devastating encounter with Uranus Minor. The volcanic and mountainous rim around the Pacific Ocean was created in this same catastrophe. The most prominent astroblemes were also products of the Age of Urania, but not all of them were blasts accompanying the ejection of the Moon.

Areas were set on fire. Elsewhere, newly created basins were being paved with basalt. As floods descended from the high land, they were vaporized on the hot lava. The water recirculated. Torrents of rain fell upon cooler land, producing another flood, which, descending, was once more vaporized, and assembled to launch yet another torrent. The sky remained cloudy while the oceans formed.

With extra charge on the globe, the Earth's volume increased. The continental blocks fractured and sometimes folded as they conformed to the underlying shape of the Earth's body, and occasionally as they underwent collision. Thus the geography of the modern world was established: separated continents, ocean basins, global fractures and ridges, mountainous ramparts around the Pacific Basin and a global-circling welt (the old rotational equator), which forms the other great mountain chain of this planet.

The => *Mohorovicic* discontinuity, which is found beneath the crust throughout the world, marks the level at which the ancient crust was severed from the underlying solid. The continents continue to float, moving today ever so slowly, but only eleven and one-half thousand years ago that motion was initiated in hours and rapidly completed. Three thousand years later the continents were almost at rest and located close to where they are now found. They came to a halt because an electrical equilibrium had been established among them, at both their prows and their sterns (see Harrison, 1966). The ocean basins by then held water roughly to the base of the continental shelves. Many seamounts (and present oceanic islands) were exposed and acquired biospheres in time. But another deluge, the Saturnian (or Noachian), was to come.

The sculpting of the ocean basins occupied a millennium. During all of this time, waters continued to descend onto Earth from the sky in rains and occasional small deluges. Both old and new waters traversed the continental masses in the gorges of the major fractures, converting them into rivers that poured over the continental shelves and into the abysses, forming slopes, too, from the large amounts of detritus that they transported. The slopes were largely formed from broad-sheeted run-offs from the continental blocks.

Within the basin, the heavy heat from continuous mantle extrusions evaporated the waters, forming dense clouds that filled and rose above the abysses into the atmosphere above the continental blocks. The early continents, until eroded, were large buttes surrounded by the new paved basins located five thousand meters below the surviving land masses.

There are indications that the drop of five kilometers into the abyss from the continental shelves was known to the ancients. Ouranos, the

Greek Super Uranus, cast his rebellious sons into Tartarus, "a gloomy place in the Underworld, which lies as far distant from the Earth as the Earth does from the sky; it would take a falling anvil nine days to reach its bottom" (Graves, Hesiod a). One notes the gloom (the dense clouds below the habitable plateaux), the position (below the human world), the precipitousness (the metaphor of an unimpeded falling object). Ancient sailors spoke of falling off the edge of the world (a fear also present in the modern child). The concept of hell, which John Locke said was so persistent that it must have represented some human experience, may have arisen from the era of the great chasms; hell is straight down, is burning, is fulminating, is sulphurous. There was, too, a world of the antipodes that could not be reached, possibly across the abyss (Dreyer, pages: 7, 37, 213, 220).

At higher altitudes on the plateaux and where the edges of the abysses were remote, snow fell and glaciers formed. Such appears to be the only scheme by which the heat needed to raise the waters can be supplied, while an area that can support ice caps may exist to receive the waters. The old ice-falls had been melted in the lunar eruption; the new ice persisted until the basins were filled up to the continental margins, and the ocean cooled; then, in the "Golden Age of Saturn", the ice melted into the ample basins. Not until the Super Saturn nova was there another "Ice Age".

The disposition of the atmosphere is a crucial problem for our model. The atmosphere would have been sucked up under a gravitational model, and unquestionably much atmosphere, and also water, was lost in the eruption into space. However, the electrosphere was already operative, as indicated earlier, and was ionizing as well as electrically repulsing, and hence returning, the gases that sought to leave the Earth.

The reduction in atmospheric pressure was short-term but unquestionably fatal in a great many instances. Even at the antipode of the catastrophe, the air would have rushed towards the scene of the disaster. Some help would have been derived from counter-winds electrically repelled and driven to the antipode. The electrosphere, containing a mixture similar to, but richer than that of the weakened plenum, would have originated downdraughts at the antipode. Assuming that the pre-Lunarian atmosphere was three times the present density at sea level and taking as the short-term extreme the habitat of people in the High Andes today, the atmospheric pressure might have been reduced to one-sixth for a short time (see Gray, pp. 63ff, White, p. 763). This would not eradicate life.

The departing Uranus Minor is deflected slightly from its path by the

Earth. It then crosses the binary axis so as to approach the Sun on the forward side (orbiting directly). Its passage by the Sun causes an electrical transaction which increases Uranus Minor's negative electric charge and ejects it into an orbit beyond the surviving binary pair. It now becomes the planet we know as Uranus, or possibly the planet Neptune (see ahead to Chapter Fourteen, p. 165, fn. 94).

The granite and mantle material removed in the passage of Uranus Minor past the Earth is strewn along an arc between the retreating intruder and the gashed Earth, there mingling with ejects from Super Uranus and travelling with Uranus Minor. A portion of this debris escapes with Uranus Minor, but most of it, amounting to about one-fiftieth of the Earth's volume, is left behind in the space near Earth. For a time some of it fell back upon the Earth as stone and dust. The rest, partly molten, was assembled by electrical pressure into a rapidly cooling globule. The form of fission of a body in such a manner was foreseen by G. Darwin and Fisher; later Baker (1954, p. 20) constructed a simple instrument depicting the process: his drawing of the critical stage of the fission is reproduced in Figure 31.

Figure 30
Fragmentation of Super Uranus

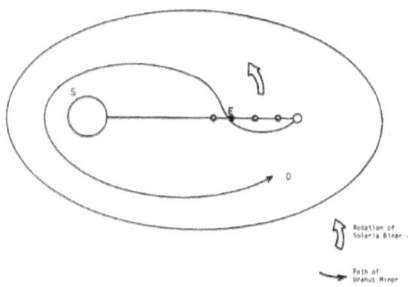

In schematic form the relocation of Uranus Minor is shown after its explosive ejection from the solar companion, which to that moment had been Super Uranus but then had become the smaller Super Saturn. The released fragment, Uranus Minor, first traveled sunwards along the magnetic tube, where it passed close by the Earth (E), tearing crust away and forming the Moon in the process. Thereafter, the still electron-rich Uranus Minor moved into the vicinity of the Sun (S) before escaping into the outer regions of the system beyond the orbit of the new companion (O), where it is likely located today as one or the other of the two most distant major planets.

A new sky god/goddess, the Moon, is born, child of Mother Earth, Aphrodite-Urania. She is worshipped after her father retires from Earth's view.

Figure 31
Fission of the Earth-Moon Pair

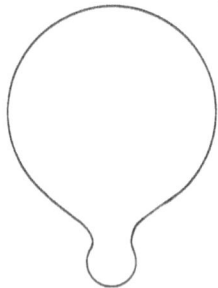

This simple diagram illustrates the minuteness of the Moon compared to the Earth's bulk, despite its having removed half of the Earth's crust when it departed.

Eventually the Moon orbited the electrical axis, repelled by its excessive charge to a greater distance from the axis than the Earth. From its removal as a piece of the Earth to the present, the Moon can never have been free of the Earth - if it had escaped it would now be an inner planet of the Sun, on an independent orbit and far from the Earth. This leads us to conclude that so long as the Earth remained in the magnetic tube, the Moon remained close by. It orbited then, as it does now, with the Earth. Though considerably closer to Earth than today, seen only in a daylight sky, the Moon was not a significant object in the sky. Likely it was larger than the disc of Super Saturn but it was incomparably fainter. It showed no phases, nor could it eclipse any body;[91] it probably was always oppositely positioned in the sky to the arc and at => *quadrature* with both of the brighter stellar bodies.

In the period of accretion, debris and lightning would be striking the Moon from the plane of the swath in large part. Far-flung cultures portray the goddess of the Moon as a spinner, the first spinner (de Grazia, 1981). In ancient spinning, as in its modern survivals, threads are held and fed from the one hand to the spindle held by the other hand (Suhr). The spindle grows fat and round. Such myths may represent the accretion of

---

[91] In support of our notion that the Moon did not orbit the Earth monthly we note that Vertes has criticized effectively the notion that certain Upper Paleolithic artifacts were lunar calendars.

the Moon and its assumption of a globular shape (see Baker, 1954, p18). The process would have endured for generations and would have been most impressive at first, especially while the Moon was candescent. Afterwards, it may have been visible only on occasion during electrical discharges.

Tresman and O'Gheoghan quote *Midrashim* to the effect that the Moon fell, was less brilliant, "and tiny threads were loosed from her body". Also, "some of her parts fell off."

The legendary evidence of the birth of the Moon is discussed elsewhere (Darwin p510; Fisher; Bellamy, 1936, pp. 268-72, 1951, ch. 16; Baker, 1954; and de Grazia, 1981). Physical and astronomical evidence is abundant on the manner and recency of the Earth's parturition and the birth of the Moon. Here we cite only salient examples from several scientific disciplines.

The Moon's overall density approximates the density of Earth's mantle material; so it is natural that many scientists have suggested some connection between the Moon's origin and the Earth's missing crust. The similarity between the chemical composition of the surface rocks of the Moon and the Earth enhances the believability of this hypothesis, especially with regard to the amount of the noble metals - gold, platinum, nickel, etc. (O'Keefe, 1973). That the Moon rocks are more impoverished in the volatile elements (zinc, cadmium, lead etc.) than Earth rocks indicates that the Moon material has been subjected to more heating than has Earth material. The Moon rocks were formed under reducing conditions: ferrous iron is common on the Moon, while it is rare on the Earth, where oxidized ferric iron is found (Arnold). If the Moon agglomerates electrically in a depleted plenum all of these differences are explained. The lunar material is heated as it is wrenched off the Earth. In space it is dispersed into an oxygen-poor dilute gas, where it accretes electrically again, liberating heat. The forming Moon is unable to conduct this heat away efficiently (as had the Earth, which accreted in a much denser and more electrified plenum at an earlier time).

The Moon's internal structure testifies to the rapidity of its formation (Wood). Despite a high surface heat flow, the Moon's interior is relatively cool today (below 1300 K); the Moon seemingly accreted as a conglomerate, like stew chunks in sauce (de Grazia, 1981). Its sixty-five kilometers of anorthosite crust reveals that it was melted or, better, metamorphosed electrically, at the time of its agglomeration. Its over three trillion craters of one meter or greater in diameter (Short, p48) show that it underwent extremely heavy electrical blasting and debris bombardment after its emplacement. As a result we are not surprised that some rocks returned to Earth from the Moon show strong and fairly stable remanent magnetization (Strangway *et al.*) despite the weakness of the lunar global magnetic field. If electrical events magnetized the Moon rocks their *in-situ* magnetizations should be quite disorganized, sporadic, and of varied strengths.

Several *in-situ* observations testify that the Moon's formation was very recent. Lunar samples do not match modern theoretical expectations about primordial planet composition (Wood, pp. 71- 5). The oxygen isotope ratio in lunar samples is identical to that in samples of terrestrial oxygen (Epstein and Taylor). The amount of Helium-4 (a product of radioactive decay) found in the Moon's rocks is exceptionally low (Heymann et. al.), indicative of the Moon's youth (Cook, 1972, p18).

Cracked crystalline surface rocks show evidence of shock metamorphism and rapid cooling (Douglas *et al.*, 1970). The bombardment has been extensive and repeated, while some debris is of recent origin (Quaide *et al.*). Using the conventional time scale, recency means about one four-hundredth of the Moon's age; using our time scale it means very recently (de Grazia, 1981; Baker, 1954; Velikovsky, 1969). The thin lunar atmosphere is accumulating now; gases trapped when the Moon agglomerated are still escaping from orifices in its surface (Cook, 1972). Periodic eruptions are reported, notably in the craters Alphonsus and Plato, and within Schroter's Valley (Menzel *et al.*, p. 229; Wilkins and Moore, p. 235, p. 263). Moonquakes, frequent though weak, may be taken as evidence that equilibrium has not yet been attained within the Moon and within the Earth-Moon system (Latham). It is presently impossible, however, to distinguish which or how much of these several phenomena are attributable to the throes of the birth or result from more recent encounters between the Moon and other planetary bodies and comets.

CHAPTER FOURTEEN

THE GOLDEN AGE AND NOVA OF SUPER SATURN

The great god, Saturn, identified by many names, of many cultures, and often associated with the planet, which has been known in the most ancient times by that name, is the son of a god. In the ancient Buddhist liturgy he is called Ravisuta, or "Son of the Sun". Uranus, by his many names, does not have a father; but he is often referred to as the father of Saturn, Uranus is killed, castrated, defeated, retired, or dismissed, and usually it is the work of a Saturn Figure. The physical circumstances of his end are those that may be associated with a stellar nova, as we have described it. So, too, Saturn comes to his end in a disastrous struggle, thousands of years later, in favor of a new planetary god, Jupiter, also called by many names, and usually made the son of Saturn. There is reason to believe that the ancients, when they used the metaphor father-son to refer to sky bodies, meant the most direct and close relationship of one body to another.

Saturn came into his own as king of the gods in the period following the destruction of Super Uranus and the ejection of the Moon. We have recited only a portion of the mythic evidence for the period immediately following the rise of the moon god or goddess; there is little coherent knowledge of human societies of the time. Thousands of years were required for the reconstruction of the Earth's surface and the recovery of a biosphere; the possibilities of monumental and record-keeping cultures were low for many generations. Perhaps the Moon would receive the extensive and obsessive worship of a great god, or even *the* great god, for some three thousand years before Saturn came into his own as ruler of the gods. More likely, we think, would be an earlier determination that the Moon was fixed and captive, pallid, and earthly in origin, hence capable safely of dependent status, along with the Earth, under Saturnian rule. Saturn would have ruled therefore either from 11 500 (or from about 8 000) down to 5 700 years ago.

The concluding date of 5 700 BP is related to proto-historical times, the time of the Great Deluge of Noah in the Bible and the First Dynasty of Egypt. The date is likely to be fixed with exactness someday. A Sumerian prism, for instance, names ten kings who ruled before the Flood, and says, "then the Flood swept over the Earth. After the Flood swept over, kingship again descended from Heaven." Wiseman, the editor of the prism, adds that "there is actually a line drawn across the text to separate the postdiluvian events from those occurring before the Flood".

At this time, as the Saturnian era moved towards a close, the disc of Saturn appeared three times larger than today's Sun. Its orbit about the Sun took about sixty-four present days. Earth was still wheeling in orbit between the Sun and Saturn, such that from Earth Saturn looked about four-fifths as large as the Sun. In a time close to three present days Earth completed its circuit about the arc.

Granted an elliptical orbit similar to that for binary stars of 64-day period (note D), Saturn and the Sun would apparently expand and diminish in size (cf. Talbott, D. N., p. 181). With this movement, time-keeping would be suggestive and simple. The Earth would also enjoy two seasons, cooler and warmer, each of more than thirty days duration.

Brahma as Super Saturn absorbs, regurgitates and reabsorbs as the ages pass (Mullen, p. 15). So does Kronos, identified here with Super Saturn. As with Super Uranus, Kronos' instability made him less than an ideal father. In Greek myth, the great god Kronos (Saturn) swallowed at least five of the children born to him out of his sister-wife Rhea. She then hid her youngest, Zeus, and fed a wrapped stone to Kronos. When Zeus matured, he led a revolt that ended in the banishment of Kronos, after he had vomited up all his children alive. The interpretation here is that Super Saturn was absorbing fragments that remained from the Super Uranus debacle or ones erupted later by Super Saturn itself. Then during its instability, it exploded back a number of them, before as a nova it fissioned into four major parts, corresponding in the myth to Jupiter-Zeus, his brothers Hades and Poseidon, and Saturn Minor, the distant planet of today.

That Saturn possessed satellites was known to the ancients (Tresman and O'Gheoghan, p. 36). That Saturn was regarded as a Second Sun, a Night Sun, is shown by Cardona (1977, p. 33) and others. That it was bright is claimed by numerous ancient texts and authors, surveyed by Jastrow, Mullen, Greenberg and Sizemore, Velikovsky (1973, 1978a), Cardona (1977), Tresman and O'Gheoghan, and Talbott (1980). That it became exceedingly brilliant just before the Deluge of Noah is implied in Hebrew legends (Tresman and O'Gheoghan, quoting Ginzberg). One Hebrew source has the star "as bright as one hundred suns." That widespread cultures held Saturn, the god, responsible for the Deluge is made clear by several authors already cited. That Saturn the planet was deemed responsible for the Flood is equally plain (see especially Velikovsky, 1978a, 1979). Velikovsky appears to have been the first to claim that Saturn became a nova, an idea that he found buried in Jewish rabbinical commentaries on the Deluge. That Jupiter was a prime element in the nova and subsequent events is evidenced in many of the same places; Ea, the Akkadian Saturn, reproaches Enlil, or Jupiter, for having caused the sky-waters to fall (Mason, p. 77). Trisiras, a son of Prajapati and a saintly Saturn figure, was a three headed god with heads resembling the Sun, the Moon and the fire, which we interpret respectively as Saturn itself, the celestial crescent and the electric arc. Indra (a Hindu Jupiter) slew Trisiras with a thunderbolt, whereupon Trisiras' three heads shone with brilliant energy until they were cut off, and then flocks of birds flew out of them and "his fever left his body."

The powers that acted in the Heavens were manifested to humans amidst increasing disaster. In terror, self-abasement and pleading, man created a Uranus-Heaven religion and hoped for cosmic tranquility. Unanimously the legends of the world acknowledge the human to be an imperfect creation; some have him created more than once, as for instance, the Olmecs of Mexico; they sensed faults in their powerlessness against cosmic forces. They projected then retrojected their faults to the behavior of the gods. These notions of imperfection were indelibly imprinted. Perhaps never until the past two hundred years did humans believe that they and the natural environment might be benignly controlled through human intelligence. What may have most bothered the early humans was their inability to manage their internal psychic systems. They were interminably made anxious and self-reflective by their lack of self-control.

From their ungovernable alter egos arise the huge variety of traits and behaviors of the gods; they are artifacts of human conduct analogized to objective features of the natural environment. In every aspect of nature could be found some physiognomic and behavioral parallel with the self and with the primary human group with which the self identified. This intrinsic, practically congenital, confusion of the inner and outer worlds of mankind was a two way transaction that led humans to emulate the most extreme and complex manifestations of nature, with results upon human nature and culture that were in modern perspective often richly "constructive", but frequently "self-destructive" as well.

When the skies opened after the lunar disaster, the new great god, Super Saturn, was visible dimly, through the clouds, to Earth's inhabitants. They had already been "religious" for millennia and might readily once more identify the sky objects with human forms and actions and project their hopes and fears upon the heavenly objects newly visible. Magic, spiritualism and animism, which have been regarded sometimes as substitutes for, and predecessors of, celestial religion, were derivative accompaniments of the human preoccupation with celestial behavior; they were forms of homeopathic social medicine for the "great disease".

The first religions were in the broadest sense "monotheistic."[92] Heaven was worshipped as the active power. As we set forth earlier, from the very beginning, humans have tended towards a supreme god. The Chinese, with T'ien, may have been the most persistent in abstracting a monotheistic idea from the Heavens and using it through a succession of specifically powerful heavenly forces. The *I Ching* gives this sequence: the First Principle is Heaven (T'ien) eternally present, chaos without form; the Second Principle and First Sun, giver of time, called "The Arouser"; the Third Principle, and Second Sun, an orderer, "The Limiter." "The arouser" appears to have been both Super Uranus and Saturn, a merging of memories over time also to be found in other cultures.

---

[92] See the works of Plato, Eliade, Lang and W. M. Schmidt, contrasting with the popular views of Frazer, Tylor, Spencer and others.

Figure 32
The Chinese Craftsman God and His Paredra

Fu Hsi and Nu Kua measure the "squareness of the Earth" and the "roundness of Heaven" with their implements. The god is depicted enthroned atop a serpent-like column arm-in-arm with his mate. It is tempting to suggest that this picture illustrates the situation during the Age of Saturn, with the god-star perched stop the "fiery" electric arc, which rose above the world and faded in the distance into the golden sky.

The ancient Persians and others asserted that God created Saturn (whence Saturday) on his sixth and last day of labor, before resting on the seventh day (whence Sabbath = rest) (Cardona, 1978a, p. 34). The implication is that all things - the separation of Heaven and Earth, the other celestial objects thus revealed, the biosphere, the advent of the Moon, and mankind - were all accomplished before Saturn appeared.

Various scholar identify the Biblical Elohim as Saturn (de Santillana and von Dechend, p. 146; Tresman and O' Gheoghan; Cardona, 1973a; and others). More likely, the story adopts the designation of god employed during the Saturnian Age (as for example, it was assertedly retold by Moses in Genesis). But "Elohim" at the beginning of Genesis is behaving like the great inactive demiurge brooding over the Pangean chaos, who then becomes activated as Super Uranus in the troubled phase, and creates the world, as mankind, born on the sixth day, received and perceived the Cosmos. Whereupon a more detailed account begins, relating the Hebrew experience with Saturn as distinct from the more general, aboriginal human experience.

When the gods changed, humans bowed to the changes. This repeated behavior over thousands of years is a significant motif in religious history. The lamentations over the death of Saturn were worldwide. Because Saturn "died" in what was an historical period, although little of its civilization remains, the hysterical and obsessive mourning shows mankind affected by, not affecting, a real tragedy. Thousands of years after the death of the second sun and the end of his age, the Roman government was acting to suppress infant sacrifice to Saturn. The parallels between Saturn and Christ as a Saturnian figure are numerous: the passion of Christ is historically and psychologically a re-enactment of the character, the unjust death, and the resurrection of the god who had died some four thousand years earlier as Osiris-Saturn. Frequent efforts philosophically to cover over the deep trench of tradition connecting the two gods have failed to divert the mainstream. Such efforts have built a distinctive existential character for Christ. The => *Age of Saturn* in cultural terms was probably what is usually designated as upper Paleolithic and Neolithic. It would be the age of Atlantis and other civilizations lost to view in the disasters that followed.

Saturn, who was generally accredited with bringing agriculture and other useful arts to mankind, was the first Lord of the Mill, a sky wheel grinding out material and spilling it upon the Earth - gold, salt, sand, and stones. Before it sank in a cosmic maelstrom, it ground salt into the sea (de Santillana and von Dechend). The Saturnian Deluge was caused by a salt-water tree cut down by a tapir, according to the Cuna Indians *(ibid)*. In Hindu myth, the gods were churning the heavenly waters and ground salt into the seas (Tresman and O'Gheoghan, p. 39, and see Figure 33).

Salt domes are among the most common of the mineral intrusions that are scattered over the Earth's surface. Small ones are about one cubic kilometer in extent, but some are several hundred times this extent. The largest known salt dome is in Calliou Island Bay, Marchland Louisiana; its size is unknown but it is estimated to be 43 kilometers long, 20 kilometers wide and 6 kilometers deep; if so, it contains about 10^{13} tons of salt. Globally 950 large salt domes are being used for mining. In some regions these domes are associated with petroleum and natural gas deposits. Other

Figure 33
The Churning of the Sea

Vishnu sits atop mount Mandara accompanied by his wife Lakshmi. The mountain of the world is being spun as the great snake Vasuki is pulled to and fro by the devas (grasping the snake's neck, to the left) and the asuras (holding its head, to the right). Together they churn the sea of milk, producing the liquid of immortality. Unfortunately their churning became so violent that it threatened the Earth, whereupon Vishnu, as the avatar of the turtle (seen here below the mountain), came to save the world by assuming the role as its pivot. Still the world was threatened by the heat of the churning until Indra sent the Deluge from Heaven to quench the fire.

- courtesy *Larousse Encyclopedia of Mythology*

places have sulfur deposits associated with the salt intrusions. Since most salt domes have been found scattered widely from the three major salt-dome fields known today, it is not unreasonable to think that only a small fraction of the buried salt has been discovered. These immense deposits of salt in the ground suggest a non-marine source of all salt. The salt in the seas can be explained in terms of salt falls from space.

Whence comes "The Golden Age" of Saturn? Saturn is golden. So is the light on Earth. The interruptions of absorbing the children were spaced out, perhaps half a dozen marking disasters over the 2 300 years that followed the lunar period. Meanwhile life on Earth may have been easy in most places. There was no ice age. Travel by boat was easy, for the breezes were mild. Antarctic may have been mapped in this age, since an ancient map showing its outline beneath the present snow apparently has been found and since no later age would have been able to produce it because the coastline was invisible (Hapgood, 1966). The northernmost and southernmost regions were quite habitable, even tropical. The continental shelves and slopes had become livable. There was a plentitude of moisture and all-year warmth. The plenum and electrosphere were still insulating. Further, Earth was holding its own surface atmosphere despite the thinning of the plenum under Saturn. The plenum became increasingly more transparent.

The arc was less brilliant and more intermittent, yet the fire was there, binding Earth to its great god. Kronos is addressed in an Orphic Hymn as "you who hold the indestructible bond," while his Babylonian alter ego held "the bond of heaven and earth" (Tresman and O'Gheoghan, p. 36). The "central fire" is also represented, likely, in the innumerable pillars, phalluses, bonds, pinnacles, pyramids, and one-legged gods (Talbott, D. N., pages: 190, 368, 188). Even the jagged sickle, with which Kronos castrated Uranus, may have symbolized the electrical axis being severed from Super Uranus. Manu, a Hindu creator god of the Flood, stood on one leg for thousands of years, contemplating his world design. All of these images are closely connected with Saturn, often by name as well as by symbol. Yahweh is imagined in Jewish legend as radiant atop Mount Zion, and Kronos-Saturn ruled Mount Olympus before Zeus toppled him.

Super Saturn sat atop the sky, a dull red disc three times the size of the present Sun. Because of the Earth's offset from the arc, the sub-solar position on Saturn's face was askew 21 per cent from the center of its disc. The arc impinged upon about five per cent of Saturn's face. Below Saturn it widened like a tree until it passed Earth's horizon, where it was something like 15 degrees wide.

The insistent worldwide legendary connections between the Pleiades, the Deluge, and Saturn as god and planet (Cardona, 1978b) point to the likelihood of the Deluge as having occurred at the time when Super Saturn had masked this star group, and of the Pleiades having emerged for the first observable time in the sky at the zenith following the clearing of this place by the actual bodies and the debris of the nova. The Pleiades were widely taken to be the remnants of the Deluge nova by the myths.

The report of Seneca, about Berossus' history, to the effect that when the stars are lined up in Capricorn, a great flood occurs, can be interpreted to mean that the Deluge occurred at the time of year when Capricorn was astrologically dominant, which, in the period when astrology crystallized, would have fallen near the end of the year. But the end of the year is the end of the Age of Saturn, hence the floods of Capricorn are associated with the Saturnian Deluge.

Also to be considered is the naming of the large sky area, "the Celestial Sea", where the two Pisces (the zodiacal one and the southern one), Cetus (Whale), Eridanus (Styx, the river), Capricornus, and Aquarius are the dominant constellations. These aquatic images suggest the presence of vast celestial waters, and their one-time general location. Celestial aquatic motifs are common, as in the Golspie Stone discovered near the small town of that name in northern Scotland (Figure 34).

At the moment of the nova the electric arc was interupted long enough to free Mars, Earth, "Apollo" and Mercury from their million-year captivity along the axis of the binary partners. Thereupon they orbited the Sun independently for the first time, moving along a plane close to that of the old binary. Their new, roughly co-planar orbits, were similar to, but much more closely spaced than the orbits of these same planets today.[93] At the time of its nova Super Saturn broke into at least three major fragments; these pieces, constituting the present Jupiter, Saturn and Neptune, all receded from the Sun following the fission. Jupiter, the largest surviving piece, took up a position near the Earth's present orbit; Saturn Minor, the intermediate-sized part, blew into space beyond Mars present orbit; Neptune receded beyond Jupiter's present orbit after depositing much water into the Earth's electrosphere.[94] In the nova, charge was expelled from Super Saturn into the plenum and dispersed in surrounding space; it was as if the mass of the system had been reduced two-and-one-half-fold; so, the planetary motions slowed considerably.

---

[93] See ahead, Chapter Fifteen, p. 174, for the fate of Apollo.

[94] The evidence is slightly in favor of Uranus Minor being the modern planet Uranus, and of the god Neptune-Poseidon being the modern planet called Neptune, a marvelous coincidence, if true. Should it turn out that Hades is the modern planet Pluto we would have to consider an unconscious mechanism at work in the naming of these "discovered" planets.

Neptune is the Latin identity of the Greek god Poseidon, brother of Jupiter, who with another brother Hades, helped Jupiter overturn their father in the nova revolt. Poseidon then was granted sovereignty of the seas and assumed his role on Earth. Before him there had been Tethys, goddess of the sea on Earth, and Okeanos, god of the celestial sea girdling Earth. By implication we conclude that Poseidon played a role in the deluging of the Earth.

Figure 34
The Golspie Stone

An interesting collection of ancient celestial motifs. We find aquatic creatures associated with the Flood which ended the Age of Saturn; the destroyer god, who felled Saturn from his perch atop the column at the center of the world; the two stars connected, representing the earlier state of the world; and the intertwined serpents, or the electric arc, which was quenched by the Deluge.

Synchrotron radiation emitted by the planets Jupiter, Saturn and Uranus has been detected and cosmic ray sources have now been associated with these planets. Saturn, like Jupiter, emits much more energy that it receives from the Sun (Milton, 1978; Hunt and Burgess).[95] The heat excess given off by the gaseous planets is an indication of their electrical nature. The emission of X-rays by a stellar source is often taken by astronomers as an indication of a very recent thermonuclear nova. Following our theory, the X-rays emitted by stars and these gaseous planets of the Solar System come from electrical transactions.

The effects of the deluges of Saturn are a subject beyond the scope of this book (see Patton, pp. 51-64; Frazer, 1916). Judging by the civilizations reported to have been inundated, the fall of waters must have been worldwide and extremely heavy. Those cultures that disappeared beneath the waters (de Grazia, 1981) are presumed to have been located at the low-lying coasts of continents on what are now the continental shelves and slopes. The disappeared civilizations are not the only clues to how much water was involved. The great river canyons that course down the continental slopes to the abyss were in existence before the Deluge and were now inundated and probably greatly eroded, as were all of the valleys on higher land.

The Great Deluge would thus top up the ocean basins of the globe, covering the continental margins left dry after the Uranian deluges. Not only would more water fall on the Earth than in the earlier cataclysms, but it would fall in a much shorter period. If indeed the water came down as the Hebrew Book of Genesis reports, in forty days and forty nights,[96] then fifty-one per cent of the Earth's water descended in three and one-half million seconds: 411 tons per square kilometer-second, 1.5 meters of rain each hour. The whole Deluge would have amounted to a 1.42-kilometer depth of rain upon the Earth's entire surface. Much of this water would have to drain into the basins in order to deepen the seas by more than two kilometers. That such a Deluge has dominated the myths and legends of the survivors is understandable.[97]

Saturn is reputed to have lit up brilliantly and preserved its light for seven days before the Deluge hit the Earth. This may be interpreted as a

---

[95] The thermal state of the inner planets is much less clear. There, the radiation balance is not sufficiently measured to allow any unequivocal statements about the presence or absence of a thermal excess.

[96] The "day" taken here is 86 164 seconds (24=> sidereal hours), but would have been different in those times.

[97] The present atmosphere contains 4×10^{13} tons of water, about 10 000 tons over each square kilometer of Earth's surface. To hold the Deluge waters the Earth's entire electrosphere must have been involved. Given its immense volume, each cubic kilometer of it was still required to hold 637 tons of water and precipitate it at the rate of 184 grams each second. An inveterate bather might measure the rains by standing under the bathroom shower for 40 days and nights.

set of pre-fission flare-ups climaxing with the nova that destroyed Saturn. If the Earth were still only 14 Gm from Saturn, the debris expanding away at 200 km/s would encounter Earth in a little more than 19 hours. If the 40-days/nights period were of present duration, the problem of depositing so much water on the Earth would be practically impossible. Millions of heavy cyclones would be needed, even one per 30 square kilometers all over the Earth. If they were concentrated at the poles by the electrosphere high above the Earth and funneled down there, the damage might be less. But then the tremendous erosion would be visible today. We think that a longer span of time may have been required, and "a day" was a translated memory of a longer regular interval unknown to us, a magical cipher, or a historical error.

The problem of deluging the Earth is nearly as difficult to cope with as the recent eruption of the Moon from the Earth. In both, almost unimaginable physical phenomena must be conjectured. The biospheric aspect, which often comes first to mind, can be rationalized on the equally incredible capacity of living populations to renew themselves. Even postulating Manu and his tiny crew or Noah and his family as sole survivors, a thousand years of exponential growth could fill the land to overflowing. As with the deluges, one does not have to move far from the extremities of legend to enter the realms of the possible.

The astrophysical aspect is more intimidating (see Kofahr). To launch the waters and other debris from Saturn is readily conceivable, given the nova. To guide it and land it requires the invention of low-probability solutions. Even if the dynamics thus far presented can be accepted with respect to Earth, how does one explain the absence of water on Mercury, Moon, and Mars, all of which would have been in or near the rush of water?[98] If they were inundated, where has the water gone? There is almost no sign of water on them, or its having been on them in oceans. Where, too, is the salt?

Two types of probability occur. The two planets may have burst out of the magnetic tube ahead of the Flood churning down towards the Sun, whereas the Earth was entrapped. Or else, the Flood can have descended the tube by a passage occupied at the moment by the Earth alone. In this case, the Moon occupied a special position besides, for it was a considerable distance from the Earth towards the perimeter of the tube. In

---

[98] Juergens (1974, 1974/75) has demonstrated that the canyons and rilles observed on the Moon and Mars and sometimes accredited to deluge and fluvial erosion cannot be water features, but probably result from the passage of electrical discharge currents.

any case, the Earth would still be luckier than Mercury, or Mars, for both of these planets have ruined surfaces and no biospheres.

Three different Jewish legendary statements refer to a diminution of the Moon in size (Tresman and O'Gheoghan). This would occur presumably after the Deluge, when the Moon followed the Earth out of the old magnetic tube and was repelled by Earth into a larger, but still captive, orbit. Two legends imply that the stars multiplied then, an expected improvement of visibility in the star-system sac. The age of Jupiter, now upon the world, introduced mankind to the light of the stars in the darkness of the night.

CHAPTER FIFTEEN

THE JUPITER ORDER

When Thor, the Scandinavian Jupiter, went into battle, and

> would grasp the handle of his terrible weapon, the thunderbolt or electric hammer, he was obliged to put on his iron gauntlets. He also wears a magical belt known as the "girdle *of strength"*, which, whenever girded about his person, greatly augments his celestial power. He rides upon a car drawn by two rams with silver bridles, and a wreath of stars encircles his awful brow. His chariot has a pointed iron pole, and the spark-scattering wheels continually roll over rumbling thunderclouds. He hurls his hammer with resistless force against the frost giants, whom he dissolves and annihilates. When he repairs to the Urdhar-fountain, where the gods meet in conclave to decide the destinies of humanity, he alone goes on foot, the rest of the deities being mounted. He walks, for fear that in crossing Bifröst [the rainbow], the many-hued Aesir-bridge, he might set it on fire with his thundercar, at the same time causing the Urdhar water to boil (Blavatsky).

The numerous electrical aspects of the god are here apparent; "the euhemerization of electricity," Blavatsky calls him. Lightning is handled by a number of gods in the history of religion, but all together these are insignificant compared with the references accorded Homer's "Jupiter the Thunderbolter" alone. It is natural to see in this literature an exaggeration of ordinary lightning strokes, but we have already stressed in earlier chapters the cosmic role of lightning-like discharges. We see in the Universe countless instances of stellar and interstellar binary currents produced by the discharge of accumulated electrical charges. Also, at any given time there are several million electrical discharges in the photospheric region of the Sun, each about one or two thousand kilometers long and lasting ten minutes (Crew). "Mega-lightning" of hitherto unappreciated voltage has been observed by satellites in the Earth's upper atmosphere, but has not been detected yet by ground observation (Turman).

When legend reports the electrical activity of Jupiter the god, it tells of the electronics of the planet Jupiter. Mountains are leveled or melted, sky monsters felled, citadels destroyed, the Earth scorched, and armies sent fleeing: all the work of the king of gods. Every lightning stroke to Earth becomes a theophany, as in lightning-obsessed Etruria, which gave the name Jupiter (Jove-pater) to the Romans. The sacred manifestations consecrate the cosmic bolts that were memorialized and discussed for thousands of years.

The planets, following the interruption of the magnetic tube, were freed. Instead of wheeling with Jupiter, now the binary component, they orbited the Sun independently, their motion close to the plane of the old binary - now the plane of the reconstituted Solar System. In their free orbits the planets avoided one another and Jupiter because of their electric charges, which produced repulsive forces when they came into proximity. Regularly they passed through, or close to, the axis between Jupiter and the Sun. Then Jovian thunderbolts were experienced. These could be the now occasional visible discharges of the dying binary, catalyzed by the presence of a charged planet in the path of the discharge; more likely they were locally generated discharges between the planet and its electrosphere, induced by the planet's voyage through the electrified region within the invisible arc-discharge between Jupiter and the Sun.[99] Either way the planet was zapped by Jupiter as it came into opposition with the Sun. From the Earth, for the first time humans might see the other planets swinging on their journeys around the Sun.

Planet Jupiter, now viewed as Ruler of the Heavens, struggled to restore and maintain the arc - connection to the Sun - for a time the arc flared with occasional visible spurts, but mostly the electric connection was dark. It is in this era, possibly, that the existence of a Counter-Earth was proposed, a dark body which obscured the celestial fire (see behind to Chapter Six).

Jupiter is the most phallic of the great gods. The association of electrical

---

[99] Today, when planets pass one another in orbit geomagnetic disturbances are noted (Jacobs and Atkinson).

stimulation, phallicism, and thunderbolting is strongly linked to the religious rites in vogue at the time of Jove (Ziegler, pp. 65-72). Phallic worship is common among Jupiter-type deities (Tresman and O'Gheoghan). The Amun temples in Egypt are liberally decorated with images of the ithyphallic god Min. Shiva (the Hindu equivalent of Jupiter) emasculated himself when the realized that his creative ability had left him. The analogy of this legend with the end of the visible electric arc is plain.

The Golden Age of Saturn contrasts both culturally and physically with the bright harsh Age of Jupiter. We must explain brighter skies, a worsened climate, a larger role for sporadic electrical phenomena, and certain striking astronomical movements of Jupiter's "Olympian family."

The Earth emerged from the magnetic tube following the Saturnian Deluge (about 5700 BP) with its rotational axis forcibly relocated.

While in the tube, it was constrained to maintain a magnetic axis along the tube's perimeter. Freed from the tube, the magnetic axis found a new alignment in the magnetic field induced by the apparent motion of the charged Sun about the Earth. This magnetism, albeit weak, established a new rotational pole on the Earth close to, if not coincident with the Earth's magnetic pole (see Lapointe *et al.*).

A small tilt and a relatively diffuse plenum made the variations in such sunlight as was released very noticeable on the Earth in the altered system. Seasonal differences in the earlier era were minimal compared to variations in climate now existent on the Earth and during the year. We have already suggested that the first lines of Genesis move quickly, and possibly in a confused way, from a Uranian beginning into the Age of Saturn. Similarly, the second and different creation, which follows a few verses later without evident attempts to reconcile the two theories, begins in a Uranian setting, of mist without rain, and before agriculture. Man is made out of earth and placed in the luxuriant Garden of Eden, in an innocent, proto-human state of unabashed nudity and unselfconsciousness. Man gave names to every creature, and was given woman out of himself. The tree of life and knowledge, planted in the middle of the garden, and the four divided rivers of Eden, are firm symbols of Saturn, corresponding to the electrical axis and the cross-sections of the Saturn disc. (Talbott, D. N., pp. 120ff).

The wily serpent that tempts Eve and Adam is the alter ego of the tree. The couple, eating the fruit of the tree upon the serpent's persuasion, become fully human, that is, possessed of self-awareness: "Then the eyes of both were opened, and they knew that they were naked; and they sewed fig leaves together and made themselves aprons" (Genesis 3: 7). Thus they

Figure 35
Apparent Motion of the Charged Sun about the Earth

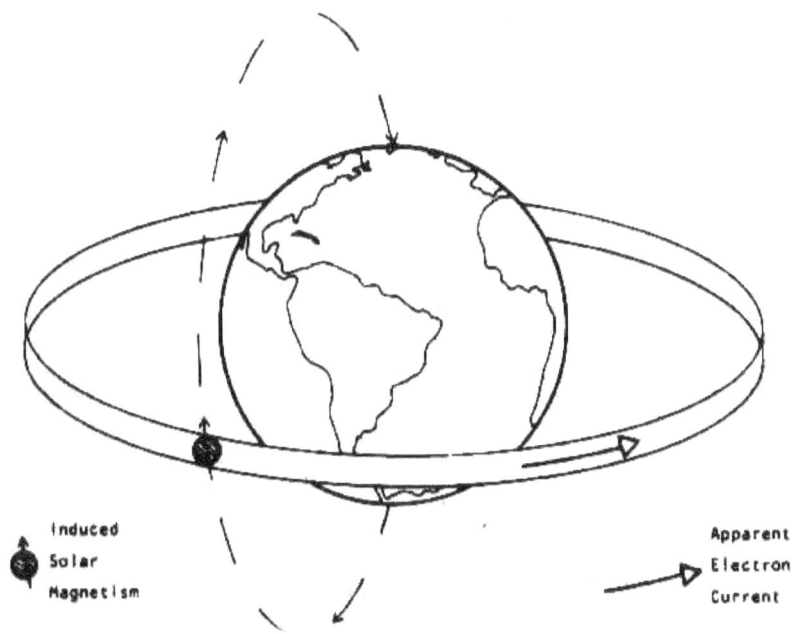

Freed from the magnetic tube at the time of the Deluge, the rotation of the magnetized Earth was brought into alignment with the weak magnetic field generated by the relative motion of the electrically charged Earth-Sun pair. From the Earth the charged Sun is seen to flow in a loop around the Earth in one year, representing an electron current flowing counter-clockwise along the ecliptic (as seen looking down on the North pole). Such an electron current creates a magnetic field, which enters the Earth at its North magnetic pole. This field parallels the Earth's magnetic field. The situation described here initially brought the Earth's magnetic and rotational poles together. The later quantavolutions separated them again and tilted the Earth's rotational axis to the ecliptic.

also work; they feel conscience already, and the wrath of the deity for what they had already become, conscientious workers. He evicts them into a world of shame and toil, far from the sacred tree (axis), the winged, lion-bodied gods (of the Saturnian symbolism) and the twisting, flaming sword (the axis again in its more visible sputtering phase). The Earth is now a drier, harsher habitat, where they had to wear skins rather than fig leaves. It is the age of Jupiter, and of Yahweh to come. The cosmic thunderbolts of Jupiter function in clear skies. There is a Jupiter Pluvius (" of the rains") but this is either one of his many powers (for he is overlord of all) or it is a reminiscence of his having played a role in provoking the Saturnian Deluge (Mason, pp. 76ff). The skies are clear because the plenum has greatly diminished in density; the heavens are transparent and the stars are seen.

A new ice age may now have begun, centered at the rotational poles. Fossil settlements of the extreme north have been uncovered that enjoyed a tropical flora (see Velikovsky, 1955, pp. 44ff). They probably date from the Saturnian age. The new ice will remain, advancing and withdrawing on occasions between and during encounters with celestial bodies, up to the present time. Today we cannot yet deduce whether the ice caps are increasing or decreasing (compare Kukla and Matthews with Gribbin, 1976; A. Brown).

The solar year under Jupiter may have had a succession of different lengths. First occurred the Saturnian year, to which we have assigned a 64-day duration.[100] Then it increased to 156 days when Jupiter receded. The Mayans possessed a 260-day sacred calendar that was central to their religious and cultural life, even while using a more modern and exact calendar (Coe, p. 9). We attribute this sacred calendar to the Jupiter-Earth synods of this era, to the time before 4 400 years ago.[101]

At the Saturnian Deluge we suspect the Earth was around 96 gigameters from the Sun. It moved outwards constantly after that to 107 Gm before the Mercury /Apollo episode of 4400 BP, then to 127 Gm as a result of that encounter. In the course of these changes, Earth's axial tilt was altered again and again[102] (see Dachille, 1963; Warlow), causing glacial retreats and advances in the extreme latitudes at each period.

The Apollo episode is most speculative. Considering the traits of Apollo, de Grazia (1984a) associated him with the asteroid belt. We see no reason to alter that finding now. After 4400 BP Jupiter orbited within the space of the "asteroid belt" we know today.

Briefly, Apollo is a great god of the Greeks. His equivalent identities are obscure: he may be Horus of the Egyptians (unless Amun-Jupiter and Horus are the same). He owns no planet in late ancient times. He is a

---

[100] The Saturnian year has been assigned cognizant of the requirement that the solar "mass" declined at the time of the Deluge (see Chapter Fourteen, p. 165).

[101] Jupiter then orbited the Sun in 390 days while the Earth orbited in 156 days, and so the Earth crossed the Jupiter-Sun axis every 260 days.

[102] The North Rotational Pole has had possibly three earlier positions, in the Yukon, in the Greenland Sea, and in Hudson's Bay (see Hapgood, 1970).

psychically remote god, and a god of plagues and remote missiles in war. He is young and a son of Jupiter. He is wise and was literally brilliant - "Shining Apollo."

The envied reputation of Apollo as the shining, "brilliant Hellenic god of peace and civilization" (see, for example, Grant, p. 1064) coincides with the idea that he was destroyed. He could not become even *Deus Otiosus*, thereby exposing the sad human experience, howbeit unconscious, that "the only good god is a dead god".

It is conceivable that "Apollo", a planet nearest to Jupiter, in the second millennium of the Jovian Age, was perturbed and then destroyed by Jupiter's thunderbolts. Apollo has solar attributes which were in late classical times exaggerated until he was often portrayed as the Sun, a most unlikely identity. The shining of Apollo, as of his brother, Hermes - Mercury (de Grazia, 1981) was most likely occasioned by flare-ups in close conjunction with Jupiter, prior to the outburst that destroyed the planet. Apollo, the god, often clashed with his father. To some (Ovenden, 1972), the asteroids look like bits of the residue of a large planet, long ago exploded.[103] The time of the "asteroidal explosion" is recent (Van Flandern) even under long-time reckoning; it is very recent if placed in the context of Greek legend. In this context, several events coincide and relate to the larger theory of Solaria Binaria.

Apollo has a younger brother, mischievous Hermes (Mercury), who is a swift, winged messenger of Zeus (Jupiter) and the gods, who is connected with electricity (especially as Thoth, in Egypt), the creator of illusions (mental problems), and is god of thieves, travelers, and healing. He, too, becomes a great god, known to many - East Indians, Mexicans, Teutons, and others. Though Yahweh reflects Jupiter, he also has qualities of Thoth; Moses was probably a devotee of Thoth, and acts towards Yahweh as Hermes towards Zeus (de Grazia, 1983a).

Astronomically, Mercury would have been next to Apollo, would have acquired atmosphere and debris from Apollo in the latter's outburst, then lost charge and would have been displaced towards the Sun. In so doing, he would have passed by Earth and Moon, inflicting considerable damage

---

[103] Nieto notes that such an explosion, which left débris estimated at up to one-tenth of an Earth mass from a planet whose bulk Ovenden assumes is 90 Earth masses, would not likely have left its débris exactly at the place which satisfies perfectly the so-called Titius-Bode "law" relating the planetary distance. This law, as we see it, is merely an expression that the planets repel one another. Nieto cites Napier and Dodd in arguing that such an event is almost impossible to reconstruct using gravitational, nuclear or chemical interactions, neither, apparently, having applied electrical theory to the problem of planetary repulsion. Had they done so, it follows that the insertion or removal of a new celestial body simply causes a compensatory adjustment in the orbits of the others.

upon both. The lobate scarps and shallowly scalloped cliffs that run for hundreds of kilometers across Mercury's face suggest shrinkage of this planet after formation (Murray, p. 42). In contrast, Earth, Moon and Mars seem to have expanded (*ibid.,* p. 41). In electrical terms, Mercury has lost charge while the other three bodies have gained it, consistent with the orbit shifts proposed in this book.

If indeed "Apollo" was destroyed, it must have been by Jupiter, which absorbed much of Apollo's material, so that a dearth of debris orbits in the space inside Jupiter's position today. Mercury seems to have escaped the full wrath of Jupiter; it was not destroyed. But it lost instead its superior orbit, beyond the Earth, and was flung much closer to the Sun. Like the Moon and Mars, it bears the marks of its devastation. Its surface is saturated with craters, strikingly similar in density of numbers to those on the Moon and Mars (Hammond). A "discrete terminal episode of bombardment" of catastrophic proportion has been proposed in an attempt to explain the similar surface destruction on these three astronomical bodies (Murray, pp. 45ff). Though some of these craters were caused by impacting bodies, especially during interplanetary encounters, they were in the main the result of electrical bombardment. The thick clusters of craters found even in heavily cratered terrains (Oberbeck *et al.,* p. 1697) bespeak genesis by electrical rather than heavy-body impact. The crater lumps noted at the site of the lunar rays on the face of Mare Cogitum were the earliest for which a bombardment hypothesis would no longer avail (Lear, p. 43, p. 38). Yet, besides Juergens (1974/75, II. 28ff), only Pickering has forwarded an electrical explanation for cratering.

During changes in orbit, electrical transactions on an enhanced level are induced. Unless a body is protected by an extensive atmosphere, and today none of these are, surface damage will result whenever electrical currents flow to or from them (Juergens, 1974, I. 21-3). Too, if the transactions are of great intensity, even the presence of an atmosphere will not guarantee immunity. When Mercury moved inward past the Earth it was severely damaged, both by its change in orbit and by its direct transaction with the bodies it passed.

Even Mercury's present orbit is a mystery. According to gravitational-tidal theory the planet's axial rotation should long ago have been locked to give Mercury one hemisphere in perpetual daylight, the other in darkness.[104] The discovery that Mercury rotates three times over two orbits

---

[104] The great eccentricity of Mercury's orbit would at best have the planet waggling, or librating much more that the Moon does as it orbits the Earth.

of the Sun has evoked remarks like "this is amazing" (Asimov) and has led theoreticians to postulate that the planet has been in orbit in its present position for less than six hundred thousand years (Gold, 1965). The state of astronomical and geological time-reckoning is such that six thousand may be read in place of the longer time (de Grazia, 1981, ch. 3).

Jupiter, like his father and grandfather, became a kind of *deus otiosus*, already majestic and less active in the Homeric Wars of Troy. There was no longer a close presence; philosophy and literature might usurp the regions of near space with abstract principles and metaphors. But among scientists, today, Jupiter has suddenly recovered some of his legendary features. Astronomers for some time have considered this planet to be a dark star (Newcombe). That it radiates considerably more energy than it receives as sunlight has more recently led to speculation that it is a yet-to-be-born star. Both views keep alive Jupiter's stellar nature long after it has ceased to be visibly stellar.

Today the clouds above the surface of Jupiter are very cold (150 K) yet the planet is very active electrically (Sutton, Gurnett *et al.*). Jupiter's "magnetosphere" is enormous: if it were optically visible, its size, viewed from the vantage-point of Earth's orbit, would be comparable to the disc of the full moon. The ion and electron currents detected within this magnetosphere represent radiation levels which would be fatal to humans (Panagakos and Waller, 1974, pp. 15ff). The radio noises generated within this region are received at the Earth, as are "cosmic rays" (mainly protons) of Jovian origin. Jupiter is, so far, the most demonstrably electrical of the planets. Jupiter is like a miniature Solar System, with its planet-sized Galilean satellites, its asteroid-sized satellite family and its entourage of comets. Everywhere the electrical imprint is there, and not always just by implication.

The three inner Galilean satellites, Io (resembling Earth's Moon in size), Europa (about nine-tenths of the Moon's size), and Ganymede (eight percent larger than Mercury) orbit in 1: 2: 4 resonance. When any two meet on one side of Jupiter, the third is located oppositely behind Jupiter or at quadrature to the pair (see Peale *et al.*). The resonance is seen less clearly in the motion of the fourth satellite, Callisto (slightly smaller than Mercury).

The surface of each of these bodies is distinctive (see Smith, B. A. *et al.*, pp. 934ff). Io, seemingly, is close to being molten. It lacks craters, but shows over a dozen caldera-like scars, which were likened to active volcanoes when an eruption was observed during the fly-by of the Voyager 1 spacecraft (Morabito *et al.*). An electrical flux-tube through which a current of millions of amperes flows between Io and Jupiter (Stone and

Lane, p. 947) has been linked to Io's eruptions (Gold, 1979). The Voyager 1 spacecraft was aimed at this flux-tube (Krimigis *et al.*) as it encountered Io. It missed the tube by seven megameters in what was labeled a navigational error; but more properly, in our opinion, the cause of the miss was an electrical perturbation (here a repulsion of the spacecraft by the tube).

The persistent connection, by the flux-tube, between Jupiter and its satellite, Io, is one of the last sites of cosmic thunderbolting between Solar System bodies. That discharges frequently pass between these two bodies has been known for several years since the advent of the radio telescope, when strong radio bursts which correlate with Io's position about Jupiter were detected (Dulk, p. 1588).[105] That a passing spacecraft was located advantageously to photograph the flash of one of these discharges was happy happenstance. The glow was interpreted by some experts as evidence of volcanism. Apparently, to think that we have witnessed directly the fire of the gods, a cosmic discharge, would seem to be too frightening (Juergens, 1980, p. 74).

Jupiter discharges only to Io today, but its repertoire and gamut may have been more extensive not too many centuries ago. Photographs of Europa show it to have lobate scarps resembling those on Mercury (Smith, B. A. *et al.*). Perhaps Jupiter zapped it, causing it to shrink upon loss of charge. Callisto, the outermost of the four, is one of the most cratered objects in the Solar System *(ibid)*, likely the result of thunderbolts striking it. Ganymede, the second closest of the four, shows a banded surface, pocked with ancient craters, then overlaid with younger bright-rayed craters, which stand out prominently *(ibid.)*: distinct electrical scars resembling the rayed craters of Earth's Moon (also seen on Mercury), which Juergens (1974/75, II. 28ff) ascribes to cathode behavior when interplanetary discharges occur. Additional description of Jupiter's electrical nature, especially as it affects the asteroids and comets, has been afforded by Milton (1982). The present behavior of the dark remnant of Super Uranus is, in sum, fully in keeping with the pure electrical theory of the Solar System and the historical reconstruction of Solaria Binaria.

---

[105] The inner three Galilean Satellites moving in resonance, as noted above, modulate the intensity of radio emission from Jupiter at wavelengths of the order of a decameter (Lebo *et al.*). Since the commensurability is probably due to electrical effects, the modulation is understood, using our model.

CHAPTER SIXTEEN

VENUS AND MARS

It is no longer fashionable to believe that Venus and Mars are beautiful bodies, with some congenital blemishes. Like Moon and Mercury, like Earth itself, they shriek of more recent disasters. The major question now is " How recent is 'recent'?" We address the question to their peculiarities of motion, position, composition, and behavior.

Myth and legend (in its deliberate attempts at science) afford voluminous material about both planets, their transactions with each other, and their encounters with Moon and Earth. Research of the past generation has evidenced that the planet Venus dominated the human cosmogonical mind in the years between 3500 and 2000 BP and that the planet Mars entered upon the competition to catastrophize the human mind in the latter 800 years of this period. Venus had hundreds of names and identities, many of them secret, sacred, and obfuscated. For example, the Hebrew word "shakris" means the Evening Star, Morning Star, to sacrifice, to kill something, to make sacred.

Logically, one initially seeks information about the first appearance of these celestial bodies; when were they born? In the case of Venus, legends of the Near East, Greece, Rome, the Teutons, the Hindus, and the Meso-Americans seem to speak of a special time of birth of a deity with a homologous syndrome of traits (Velikovsky, 1950).

Although the name "Venus" may not originate directly from "venire" (" to come"), as Cicero would have it (Lowery, Grant), it may well emerge even more significantly from *Venus,* which Bloch translates as "blooming nature", hence, as we see it, something new-born and rapidly expanding. The lotus and lily, two life forms suggestive of blooming cometary images, are employed in widely separated cultures as Venusian symbols.[106]

Clearly conformable to an astronomical operation is the birth of Greek goddess Athene, who sprang fully armed with a shout from the brow of Zeus (Hesiod b). The Hindu Devi is remarkably similar in the commotion that she causes when born (Isenberg, p. 90). Various studies analyzed by de Grazia (1981, 1982a) set the time of her birth near 3450 BP, in accordance with older studies by Velikovsky (1952, pp. 1-53, 98-101). The time coincides with the Exodus of the Hebrews from Egypt under Moses, an event so fraught with catastrophe that it remains the substratum of the Judaic, Christian and Islamic religions.

If Venus was erupted from Jupiter, it conceivably burst from the disturbed area of the Great Red Spot. Although not demonstrable, this is hypothetically feasible. It was Jupiter's greatest discharge, its last attempt to rid itself of ions and gain electrons. It succeeded; it retired; and its offspring was unleashed into the inner Solar System, where all massive fragments had gone hitherto.

Unlike Uranus Minor, Neptune, and possibly Pluto, Venus was of low electrical density and fell victim to encounters with Earth, then Mars, and remained within the inner circle of planets. Both the Earth and Mars took electrical charge from Venus, not without extensive physical "damage" to themselves; they both moved away from the Sun after their encounters with Venus (Ransom and Hoffee, Table 1).

Venus was not rich enough in electrons to be coveted by the Sun. By the legendary and historical evidence, it took hundreds of years to achieve a "safe" orbit from where it would not venture close enough to Earth to endanger it and cause electrical damage. Until then, it may have threatened the Earth about every 52 years. During a seven hundred year period both the Jews and the Meso-Americans observed a great "Jubilee year" on those occasions; the inference from the holiday is that the proto-planet, still behaving as a comet and undergoing continuous electrical transaction, was due to arrive in the vicinity of Earth but prayerfully would deign not to

---

[106] The lotus was used earlier in imagery portraying the central arc of fire arising to Saturn, which probably lent force to the Venus symbolism on the logic that the electric arc, thought to be lost, now reappeared, freed to roam the heavens.

destroy the world. It is significant, as one of numerous details in the Venus mosaic, that the American Pawnee Indians until a century ago celebrated a Venus festival on each occasion of the reappearance of Venus, sacrificing a virgin to the star. But many another celebration of Venus could be cited.

Granted that history and legend impart an aura of youth to the planet Venus, one can search for and find confirmation of youthfulness in the present state and observable behavior of the planet. We cannot do more than summarize here the debate upon the question, which has involved leading figures in Astronomy and Physics (see de Grazia *et al.* 1963; Talbott *et al.*, eds., 1976; Ransom, 1976; Goldsmith, 1977; Greenberg *et al.*, 1977, 1978; and many others referred to in de Grazia 1984d). We let the reader appraise the arguments dispassionately with the caution that theory must always bow to the demands of direct observation. We advance the following points in respect of the anomalous nature of Venus when viewed from the standard cosmogonical model, but implying the recency of Venus and its electrical nature in consonance with the thesis of this work. The 925 K surface temperature measured by landed space probes has not been explained satisfactorily (see for example. Firsoff; Velikovsky, 1978b; Forshufvud; Greenberg, L. M., 1979; Morrison) by any theory other than recent and continuing electrical transaction. The observation that the lowest thirteen kilometers of the atmosphere are glowing (Panagakos and Waller, 1979, p. 3) and that lightning occurs in the Venusian atmosphere (Taylor *et al.*) encourage us in our view that electrical currents flow between Venus and surrounding space. The two localities which were photographed show evidence of recent surface devastation. Seemingly the surface of Venus is similar to those of its neighbors even though the latter lack atmospheres (Ksanformaliti *et al.;* Florensky *et al.*).

One of the major surprises greeting the explorers of Venus and the theorists who welcomed their data was the demonstration of the slow retrograde rotation of the planet. Several non-electrical explanations have been offered to explain how Venus might have reversed its original forward rotation (Singer, 1970; Ingersoll and Dobrovolskis; Kundt). In a two-planet encounter involving electrical polarization (which induces aspherical shapes onto both bodies) strong "tidal forces" act and can alter spins, or flip planets over, as Warlow shows. The five bodies of the inner Solar System exhibit a spectrum of spins and orientation, running from fast-direct (Mars and Earth) to slow-retrograde (Venus), with Mercury and the Moon in between. Given a number of encounters among these five bodies over the past several millennia, many combinations of rotational alteration must be expected: it is probable that none of the spins are virginal.

The atmosphere of Venus presents another type of problem. Its carbon dioxide composition is like that of Mars, unstable over a long time due to photolysis by ultraviolet radiation. That most of it has not reacted with the exposed surface rocks is termed "surprising,"[107] possibly Venus' atmospheric gases have recently been modified.

The chemistry of the deep and dense Venus clouds has been inferred only by indirect evidence despite the passage of several spacecraft through them. Above the clouds, which seemingly insulate the planet, temperatures resemble those found at comparable altitudes above the Earth. Insolation and heat radiation from the clouds do not betray the hellish heat that was discovered below. Even at the base of the cloud layer (which is twenty kilometers thick), temperature, pressures and winds remain Earth-like (Burgess). It is the descent through the remaining forty-nine kilometers that confounds expectations and confuses the instruments of the descending space probes (making some of them inoperative and the data from others uninterpretable): here Venus does not resemble any environment yet penetrated by instruments. We find a murky and stagnant inferno under crushing pressures many times greater than those on Earth. There, lightning occurs frequently (Taylor *et al.*), indicating an electrical instability in excess of that found on Earth, and an electrical glow, noted above, permeates to the bottom of the murk.

Even the edge of Venus' sphere of influence has produced the unexpected. Venus is only trivially magnetic; yet its interaction with the solar wind produces 80 percent of the effect generated by the three-thousand-times more strongly magnetized Earth (Russell, C. T. *et al.*, 1979). Since Mercury, whose magnetism is similarly miniscule, also interacts strongly with the solar wind (Ness *et al.*, p. 480) we are drawn to conclude that the interface between planets and solar wind is electric, rather than magnetic. The bow wave in front of the planet and the long tail in the wake of the planet represent the junction of two electrospheres: the solar wind flows on one side of the junction and the planet driven ions and electrons flow on the other side. Electrons traversing this junction behave differently at Venus (and Mars), for there they are slowed, rather than accelerated as happens with Earth, Mercury and Jupiter (Simpson *et al.*; Wolfe *et al.*). Electron-deficient solar wind atoms seemingly penetrate and are absorbed by Venus' upper atmosphere (and Mars' surface). The magnitude of the

---

[107] On Earth the sea water contains 98% of the total potential atmospheric carbon dioxide as dissolved gas (Plass; Sundquist *et al.*). The carbonate rocks of the crust have locked away most of the potential supply of this gas.

effect indicates that Venus is farther from equilibrium with its surroundings than are the other planets. This finding may be the "Rosetta Stone," telling us why near Venus' surface the heat is infernal. Any unequivocal evidence of disequilibrium tells us that Venus is indeed young (Van Flandern).[108]

Earth's history of the period around 3 500 BP, so far as it is known, provides proof of extraterrestrial damage. In this respect, the following propositions have garnered enough probative support to be acceptable as leading hypotheses.[109]

1. Astrosphere:

the skies were disturbed and celestial motions were reportedly irregular. Astronomical alignments of before this age are out of line with references of the period following. No temple can be adduced whose orientations have remained the same through this time.

For instance, the second rupestral temple at Wadi es Seboua(s) in Upper Egypt was originally orientated before 1 500 BC from its rear, through its portal, via a faraway mountain saddle to the winter solstice bearing 30° 9' 0" South of East. It was destroyed by fire. Much later it was excavated and rebuilt. Again it was orientated to the winter solstice, this time, however, to 35° 49' 12" South of East (Roussel).

Possibly a block of the Earth shifted northward or else the axis of the Earth tilted; both are possible, and each may have contributed to the need for realignment.

2. Atmosphere:

There were radical disturbances and some lasting changes in atmospheric electricity, radioactivity, temperature, winds, climate, and solar radiance. The Book of Exodus can be read as a meteorological journal - one encounters electrostatic phenomena, gales, dense clouds, a unique darkness, falls of manna (compound manufactured in the atmosphere), dense rains of stones, fire (often apparently electric), a mass of dead quail (driven down by electrical storm and hurricane winds and said to be poisonous to eat), and radioactivity (see de Grazia, 1983a).

A sharp rise in C-14 levels occurs about now (and at the time of the Mars incursions 700 years later) (see de Grazia, 1981). It is doubtful that C-14 is a valid and reliable clock, since its formation and absorption rates are so easily altered by changed

---

[108] Van Flandern, in a letter to C. L. Ellenberger.
[109] From a paper by A. de Grazia, 1980, to the Society for Interdisciplinary Studies in London. Published in de Grazia, 1984c.

environmental conditions, and so evidence of this nature, though favorable to the hypothesis, must be discounted.

3. Geosphere:

Every geophysical process gives evidence of quantavolutionary stress. Widespread earthquakes are part of the destruction of towns to be referred to below. There occurred "a major westward shift in the Euphrates system of channels as a whole during Kassite times" (Paterson) (of this age, we believe). A set of natural disasters plunged the Harappan culture of India into a fatal decline now too.

4. Biosphere:

Unusual biological behavior occasioned by habitat disturbance and environmental stress is evidenced. The behavior of animals during the Plagues of Egypt is well known and not to be dismissed as a myth: it is typical of well-observed disaster behavior (Galanopoulos and Bacon, p. 192-9; Lane). In the Black Sea, a large belt of coccoliths at the bottom returns a 35 century-old Carbon-14 date at a level below the sea floor.[110]

5. Ekosphere:

All human settlements suffered destruction or damage from natural causes. In one study, we read: "In the middle of the second millennium BC the ancient cities of Southern Turkomenia declined and were abandoned by the inhabitants. The South Turkomenian civilization perished at about the same time as the Proto-Indian ... and the reasons are still unknown." (Kondratov, p. 164) Schaeffer's survey of some 40 most important archaeological sites in the Near East arrived at the same conclusion for the same time.[111]

6. Historisphere:

All contemporary accounts or chronologically assignable legends dealing with the period mention a general natural disaster. The prime case is the Bible. The Pallas Athene instance is also referred to. The Ipuwer papyrus has strong support now as

---

[110] But this is subject to questions concerning all C-14 dates before 500 BC (Blumer and Youngblood).

[111] Here we use the traditional Exodus date for the corresponding end of the Middle Bronze Age rather than the date, about 150 years earlier, used by Schaeffer.

an eyewitness account of the catastrophe ending the Middle Bronze Age in Egypt (Velikovsky, 1952, pp. 22-9; Greenberg, 1973; Sieff, 1976, p. 14; Bimson).

7. Anthroposphere:

Every culture-complex changed markedly. We have mentioned several major civilizations which declined sharply or fell - Egyptian, Indian, Kassite, Turkomenian, and others of the Near East. One might add the Minoan of Crete and the Chinese. Social organizations, religions and modes of life were altered. A corollary is that no god of before 3500 BP remained without change of status or serious accident, citing the advent of Athene and the Mosaic renewal of Yahweh as examples.

8. Holosphere:

In summation of the foregoing seven propositions, we may assert that all spheres of existence quantavoluted about 35 centuries ago. Nor does any sphere change independently of quantavolutions in other spheres. Since all spheres are changing, a general cause must be sought. There can be only one necessary and sufficient cause of the set of quantavolutions, which must be a very large body encountering the Earth. By observation and later commentaries, cometary behavior is indicated. Nothing but a god-like comet could have produced the phenomena of 3500 years ago.

It follows, finally, that every institution, behavioral pattern and natural setting that exists today, if its history is complete, will reveal an inheritance of effects from the (Venus) quantavolution. The de-traumatizing of the human mind by designing and propagating new models of natural and human history would appear to be a necessary preliminary to peace and progress. For the later time, about twenty-six centuries ago, the Mars case offers a similar set of propositions, although the evidence argues for a level of destruction appreciably lower than that obtained form the earlier Venus-Earth encounter .[112]

Around the solar year 2776 BP, human activities related to celestial disturbances were generated respecting Mars, as well as Venus, and are notable in the Near East and Mediterranean world (Velikovsky, 1950, p. 265ff). Enough subsequent benchmarks were provided by legends and practices for Velikovsky to surmise that a large heavenly body, apparently Mars, was threatening collision with the Earth at fifteen-year intervals. The Mars encroachments may have been initiated by Venus, which, pursuing

---

[112] This is expected because Mars is smaller than Venus. When Mars "struck" the Moon the damage was very great.

an ever-shortening orbit, perhaps encountered and displaced Mars from its earlier orbit between the Earth and the Sun (Rose, 1972).

The Romans were Mars-worshippers *par excellence,* and the legends, rites and early reports that tie Mars to the history of Rome are not to be disregarded; "archaeological and epigraphic discoveries" continue to demonstrate that "the legendary guise of the traditional material actually masks a real foundation of authentic events" (Bloch, p. 1085).

By contest with Venus, Mars seems to be an old god. Much less is made of his origins and birth in the Mediterranean world. When he bursts upon the world scene in the eighth century BC he is already well known. Least personable of the Olympian deities, Ares (Mars) is portrayed as a ruthless warrior. Hercules seems to be one of his more interesting identities. New militaristic nations, particularly the Romans, the Assyrians, and the ancestral Aztecs, forged empires under his inspiration. The Roman dedication to Mars is well known. He was believed to be father of Romulus, their founder. In the old calendar they named the first month after him.

The Romans irreconcilably claimed both Aeneas, Prince of Troy, and Romulus as their founder. Aeneas was, and is, placed in the twelfth or thirteenth century BC with the Trojan Wars, by older scholarship. Recently the Wars have been brought into later times, along with Homer, who sang of them (de Grazia, 1984a). This is but one step in a reconstruction of chronology that eliminates the several centuries of a so-called Greek Dark Age and pulls the disastrous collapse of the Mycenaean civilization down to the eighth century as well (Isaacson). Roman legend has Romulus and Remus (abandoned and miraculously suckled by a wolf in their infancy)[113] founding a town called Rome, which Romulus rules until he is lofted into the air, possibly by a cyclone, to join his father, Mars.

It seems to us reasonable that around this time, Aeneas and his band of experienced and cultivated Trojans might have impressed themselves upon, or been welcomed by, the local Latins of the new town, and perhaps even helped by their neighbors to the North, the Etruscans, who themselves were of Anatolian origins.

It is an age of destruction and movement. The powerful nearby Etruscan state was staggered by natural disasters and a decline. According to Pliny, their city of Volsinium, where stands Lake Bolsena today, was destroyed by a thunderbolt. Both Mount Vesuvius and Mount Etna underwent => *Plinian* eruptions around the same time.

---

[113] The wolf (cf. Fenris in Norse myth) is often a Mars symbol.

Many peoples were on the march or fleeing - possibly the Etruscan elite had not preceded Aeneas by long. They were from the general area of Ilium (Troy). Southern Italy and Sicily were being heavily settled by Greeks, in trouble themselves and profiting from natural disasters that were besetting the earlier inhabitants. And at that time the Dorians (Heraclids, sons of Hercules-Mars) were moving in upon the hapless Myceneans.

All of these loose connections and temporal references need a thorough analysis; indeed, we require nothing short of a total review of the most ancient history, legends, and the records of excavations. Even then the human record, like the fossil record, is scanty. One may hope for, but not expect, that the thousand-to-one chance will occur, and that a tablet or papyrus describing clearly the behavior of the planet Mars in these times will be found. The earliest Etruscan, then Roman, calendar was of ten solar months. That the later twelve months alternated at thirty and thirty-one days does not fit the present lunation, which is better suited to the 29-and 30-day alteration used in all surviving lunar calendars. The indication is that Mars disrupted the month in its transaction with the Earth and/or Moon.

The Aztec Mars was Huitzilopochtli, born of Mother Earth from a ball of humming-bird feathers that fell from the sky. His color was blue, his totem the eagle, and his weapon the blue snake. The interminable human sacrifices of the Aztecs were in his name or to the Sun in his name; the Sun, "the Eagle who rises," was believed to require the hearts of an unending stream of prisoners-of-war if he was to remain constant in behavior. The High Priest of Huitzilopochtli was called Feathered Serpent, priest of Our Lord; but "feathered serpent" is rendered Quetzalcoatl, who at a much earlier time was the ruling deity of Meso-America and was identified unfailingly as the planet Venus by many scholars.

The Assyrians of the eighth and seventh centuries, "like the Romans, those other stepchildren of Mars, and more than the Hittites, its victims, had a lively reverence for the planet Mars: 'Nergal, the almighty among the gods, fear, terror, awe-inspiring splendor,' wrote Essarhaddon, son of Sennacherib." Evidence of physical destruction by fire and earthquake is abundant everywhere in these years, and it was "beneath the exploits of Mars ... that Assyria marches to world power," Sieff (1981) declares. Patten *et al.* argue that the Assyrians timed their major offensives to coincide with cosmic approaches of Mars to profit from the physical disorder and consternation of their enemies.

In general, the same hypotheses that we stipulated earlier for the Venus encounters may be translated for and applied to the Mars encounters seven centuries earlier. That is, we can adduce some evidence from around the world of deep disturbances in the six spheres and of their interconnections in the holosphere.

Ancient astronomers and writers appear to have had no difficulty in considering (or perhaps they were really reflecting upon) historic encounters governing the planets; yet we can observe in their own times the strengthening of three psychological defense mechanisms that made historical reconstruction involving quantavolution difficult: denial and suppression of memory, religious and literary sublimation, and abstract philosophy.[114] Modern cosmogonists, sternly trained in the principles of uniformitarianism and gradualism under a very long whip of time, are loath even to consider large-body departures from presently observed motions. And it is true that the constraints on motions required by strict obedience to such physical laws as the principle of conservation of angular momentum are formidable; immense forces must be invoked from somewhere so as abruptly to alter the motion of bodies. Calling upon gravitational force, as this is presently perceived in science, requires "impossible" conditions even if an encounter seems "reasonable" to expect.

However, if the moving bodies are charged and are transacting electrically, many "surprising" and selectively violent alterations can happen. The observations by the ancients that passing celestial bodies appeared like the objects we, today, classify as comets are understandable. Juergens (1972, p. 7) has shown how comet-like behavior (and appearance) results when astronomical bodies move quickly from a region with one level of electrification into a remote region differently electrified. Milton (1980/81) has generalized Juergens explanation to apply it to other "non-gravitational" celestial motions. Encke and later astronomers have noted with surprise how cometary bodies sometimes alter their angular momentum in seemingly sporadic episodes (Sekanina). By this point in our study of Solaria Binaria, we should not have to digress further in order to establish the capabilities of electrical motions.

We can go further. The Moon has undergone its tribulations whenever the Earth has been engaged. Legends, and myths of Moon encounters with Venus and Mars, are at least as numerous as those involving Earth alone. Documentation has been presented elsewhere.

Juergens (1974, 1974/75) has analyzed reports of various disparities between surface features of Moon and Mars, and explained them as effects of anode and cathode behavior that would be excited in close encounters.

---

[114] Aristotle (Metaphysics); de Grazia, 1977, 1978, 1981, 1983a, 1984a; Stecchini; Velikovsky, 1982.

He has suggested, for instance, that some of the hill ranges of the Moon are so morphologically anomalistic as to represent a job of "electric welding" done by Mars.[115] To extract from these hills and from the canyons of Mars comparable material is presently impracticable, even though we may conceive of chemical tests to apply to the material thus obtained.

Juergens has gone further in the investigation of electrical encounters between Moon and Mars. On the assumption that the ray-surrounded crater Tycho (the most prominent feature on the Moon under high-angle lighting, that is, near Full Moon) could have been blasted out of the lunar highland rock by an electrical explosion liberating almost 10^{17} megajoules of energy and requiring a transfer of 10^{11} coulombs of charge between the Moon and Mars, Juergens sought a suitable anode site on Mars' surface which might receive this discharge. He found a likely receptor in the mountainous feature called South Spot (now, Arsia Mons). He writes,

> this spot is an enormous pit 140 kilometers across at the crest of an impressive 17 kilometer rise from the floor of the Amazonis basin to the west.

He observes that this volcano-like structure has no known counterparts on the Earth. Tycho also represents a lunar high point: it is some 1.2 kilometers above the hypothetical lunar sphere. The electrical connection between this feature and the Martian South Spot could have resembled Figure 36 just before the discharge occurred.

Juergens' analysis might well serve as a launching platform for decades of detailed studies, using a quantavolutionary electric and recent-time model of each planet's topographical peculiarities.

As with Moon and Mars, so with Venus and Mars. Perhaps, in the end, we shall come to regard a famous scene of the *Iliad* of Homer as an eyewitness account, garbled, to be sure: that is where the goddess Athene, archetype of planetary Venus, engages in awful sky combat with the god Ares, who is recognized as the planet Mars.

> Athene drove her chariot towards Ares, bane of mortals, and drove her spear deeply into his belly. Thereupon arose a huge black cloud, and he bellowed like 10 000 warriors and fled into the high heavens.

Some 2700 years later, the United States of America sends Mariner 9 into space and photographs the most prominent feature of Mars, the Coprates Canyon complex. It is a 12 megameter line of seemingly extruded

---

[115] Personal communication to A. de Grazia.

Figure 36
The Electric Field between Mars and the Moon

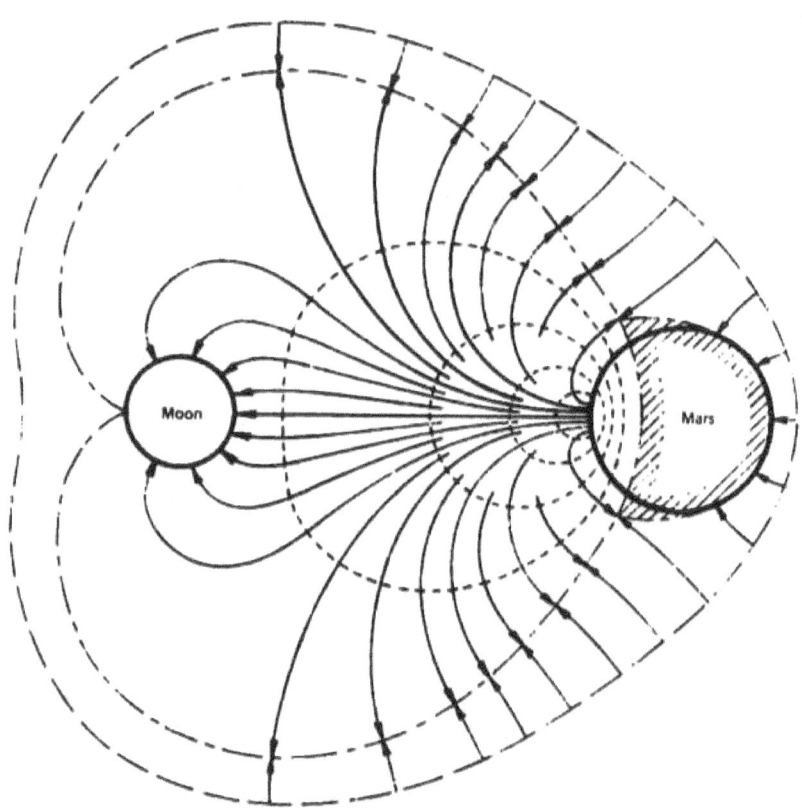

During the close passage of Mars and the Moon, *circa* 27 centuries ago, devastation was produced on the surface of both bodies, the result of interplanetary discharges across the gap between the two bodies. Here we see the electric field lines (arrows) and the lines of equal electric potential (variously dashed) prior to the breakdown that produced electric discharges from the Moon (the cathode) to Mars (the anode). The outer equipotential depicts the sheath boundary between the electrosphere of the charged bodies and the interplanetary plasma. - courtesy of the estate of Ralph Juergens

mountains, enormous beyond anything on Earth, leading into a canyon 3,600 kilometers in length, 500 km. wide, and 6,100 meters deep, ending in another string of great volcanic outbursts. Some 8.5 million cubic kilometers of rock have disappeared. The wound stretches across half the circumference of the planet.

Allan Kelly has offered a scenario. The total eruption
> is a product of the same event, when some very large comet or other massive intruder from space passed too close to Mars .... This intruder literally sucked the lava from the interior of Mars to form the huge volcanoes .... As it came closer it caused a tremendous bulge, miles high, that burst open along the top and spewed out lava and great chunks of Martian crust; much of this material followed the intruder into space.

There is no evidence of water erosion of the steep stairs of the canyon, no sedimentation, delta fans or stream beds cutting the lengthy lips of the wound. The intruder, be it Athene, Ishtar, or Venus, both repelled and implanted charge onto the defenseless Mars. Leaders of lightning from Venus incised the masses that had assembled in a belt of Mars along half its girth. In the ripping blast that followed, Mars flung them upon the intruder, and, lighter but more heavily charged, withdrew. So the aforesaid "garbling" of history lay partly in not perceiving that Mars laid a mass upon Venus and that the greatest spear thrusts of lightning were exchanged beneath the dense cloud that poured from Mars' wound.

CHAPTER SEVENTEEN

TIME, ELECTRICITY AND QUANTAVOLUTION

With the model of Solaria Binaria constructed here along the lines of short-time electrical quantavolution, we have presented physical and cultural evidence of several major historical happenings, as well as some lesser events that need not be summarized here.
1. The succession of great gods in human history coincides with a succession of ages of destruction and renewal that may tentatively be numbered at seven. These are carried in Table 6.
2. Human nature originated abruptly with a complex culture in the first age of binary instability, precipitated by electrical and hormonal changes, and displaying anxious self-awareness and a grasping for self-control.
3. The Moon was ejected from the ancient southern hemisphere (the modern Pacific Basin) later in the same period in an electrical encounter with a piece of planetary debris originating from an explosion of a star that we call Super Uranus.
4. The planets have assumed their present order in the past few thousand years, responding to a principle of mutual maximum repulsion.
5. The Solar System sac and the plenum which it contains are now

so enlarged, and hence distant from us and dilute, that they have been overlooked by observers. The binary electrical axis has been diffused into a pervasive solar wind, which permeates the planetary plane. The once-substantial binary partner is dispersed into at least a dozen sizeable fragments and myriad fragments of smaller debris. All of this has happened in fourteen millennia.

6. All major chemical and biological developments occurred in a period of a quarter of a million years at the beginning of Solaria Binaria. The number of species peaked in the period of Pangean Stability and has been steadily reduced by catastrophes.

7. The planets and the Sun are accumulating electric charge and have separated greatly, whereupon their ability to discharge (take charge) from one another is diminishing with time. If the trend continues without sudden galactic interruption, the planets will disappear upon attaining the higher level of charge found in the Galaxy surrounding us.

Contemporary cosmogony may be said to lack a binary model for solar and Earth history, and this we have attempted to provide with Solaria Binaria. We have conveyed it by means of a short time chronology, a concept of quantavolution, and a fully demanding theory of electric behavior.

Several major observations promoted consideration of the Solar System as a binary development. A growing realization that our star system is embedded in a Universe largely composed of multiple star systems led us to match known characteristics and behaviors of these systems in their varied stages of development with our own system as it might have been, is and might become. Evidence mounts, too, that planet Jupiter has stellar traits, as have, less obviously, the other major planets.

Exploration of the inner planets, not excepting the Earth and Moon, reveals the progressive destruction of their surfaces over time. Understandably , conventional cosmogony seeks to fix the destruction in a convenient episode close to the birth of the Solar System. However, the evidence speaks not of a day's work of a passing body four aeons ago, but rather of the normal work expected of a binary system.

Furthermore, ancient observers and philosophers who were neither primitive nor naive, an who were also reporting the ideas of other experts of hundreds and thousands of years before them, affirm that the bodies of the Solar System and the stars changed their behavior and their motions. These men were cognizant of, and disciplined by, religious systems that were sky-obsessed, and which moved continually between celestial behavior and mundane behavior, in supreme efforts to let happen on Earth what happened in Heaven, and *vice versa*. As we have reviewed their ideas and reports, coupled with evidence emerging from legends and archaeological excavations, we found reason to think that they might be living in a world that was strikingly different from our own and that was recognizably a late phase of a stellar binary system.

The prospect before us, then, was to understand an ancient science and tradition which had large heavenly (god-like) bodies hovering over the Earth in what today we would classify as a synchronous orbit. If the Earth were locked between the partners in a binary, much of what the ancients spoke of as their own experiences, or related as the experiences of their ancestors, makes sense.

By no means were their ideas purely deductive. They recited experiences; they made empirical statements; they claimed knowledge of a world into which the human race was born. They discussed a set of events that should occupy, by temporal schemes in vogue today, millions (if not billions) of years, and that treat matters such as the acquisition and transformation of the Moon and the recession of the planets deeper into space.

As soon as we began to draw upon ancient opinions concerning cosmic events, we had to take a position respecting chronology. In this we were encouraged by the binary concept itself to call time into question. Binary systems offer evidence of great forces operating over short times to produce large effects. They illuminate shortcuts in the creation of the raw materials for planetary, atmospheric and biological development. As they transact, the large bodies separate and fragment within the system, creating and destroying worlds while retaining their parts. It may be fair to say that only a binary model can supply those scientists - admittedly a small minority - who are inclined to shorten natural history with an adequate theoretical instrument.

The binary model suggests and may even require, a short-time scheme for natural history. A short-time Solar System requires high energy and precise interventions at levels of nature ranging from the Galaxy to the atomic nucleus. It is specifically this sort of intervention that is evidenced time and time again in natural history. All "absolute chronometries" become variable in a quantavolutionary world. The rampant inflation of this century, which has expanded the time scale for the lifetime of the Universe from 40 million to 80 billion years, may end in a catastrophic implosion.

As if their technical difficulties were not sufficient to disable long-time chronology (de Grazia, 1981), ancient human voices seem to testify against it. These earliest humans unmistakably assert, among other things, that Heaven and Earth separated, that suns appeared, that gods fought in the skies and invaded the Earth, that the world was repeatedly built and destroyed. They are neurotically obsessed with all celestial bodies and motions, and engage in all known extremes of behavior in imitation and appeasement of the behavior observed in the skies.

It is not possible to claim that this is primate activity, nor hominid, nor that it is primitive, nor finally that it is a collective psychosis of early civilizations. Modern social psychology and psychiatry can document, and even replicate, such human behavior today. The earliest cultures, those that are "guilty" of this behavior, invented social organization, agriculture, manufacturing, science, and the arts. To think that they could do all this without a firm "reality principle," as Freud has termed it, must be in error.

The skeptic of our interpretation may be reduced to postulating that *"illud tempus"* must have been exciting and stressful, but could not be so very exciting and stressful as they would have us believe. Reasoning similarly, one could assert the contrary: the real foundations of the ancient excitement and obsessions must have been even worse than we are given to believe, because the ancients were used to disasters and hence were less traumatized by them; "war is hell," but less hellish to old soldiers than to recruits. It seems to us that both *a priori* views - that the ancients were excitable or that they were blasé - may obstruct the necessary work of delineating, bit by bit, the experiences of the ancients from the conglomerated assemblage of fragmentary records, legends and geological and archaeological facies, and then exposing them to analysis in the light of the sciences today.

A persistent theme of the ancient voices is quantavolution, that is, that the world and all that is in it owe most of their changes to forceful torsions and saltations. That quantavolution plays a role in the theory of natural and social science has never been denied. But the role has been grotesquely reduced by ignoring it and stressing evolution, by consigning manifestations of it whenever possible to times beyond mind, by framing scientific principles in prejudicial terms, by associating quantavolution with disreputable or outmoded religions and scientific beliefs and by unconscious editing of the evidence.

To our view quantavolution affords an instrument for scientific inquiry as useful as and perhaps superior to that allowed us by evolution. We find that the morphology of the Earth and the patterns and compositions of the skies bespeak quantavolutions. In biology, we see in the decline of evolutionary power over time, in the absence of transitional types in evolutionary branching, in the waves of extinction of species, and in the failure of evolution to provide an internal guiding dynamic, sufficient reason to promote the concept of quantavolution.

The guiding dynamic for quantavolution, whether in biology, geology or astronomy, may be electricity, a "strong force" that has been generally accorded a weak place in most sciences. For several reasons, we believe that electricity is the necessary and sufficient impulsion of cosmology. We noticed that a strong force is needed to accomplish change, whether in biology or astrophysics. Basically, electricity is to "gravitation" (if such exists independently of electricity) as 10^{36} is to 1.

The behavior of stellar bodies, including the Sun, can be described in electrical terms. The composition of "space" is a plenum of charges and ions, field and currents, winds and relatively stationary matter, of orbiting bodies shifting orbits as they transact, at times attractively but usually repulsively.

The fact that electricity is present in all matter, and an aspect of the existence and activity of all matter, presents us with the opportunity to study all matter and motion in an electrical perspective. Electrical attraction and repulsion seem to operate simply and flexibly in cosmology as well as in microbiology, and can be accommodated to the concept of inertia, the two together constituting a powerful instrument for the analysis of nature.

Finally we would point out one more helpful attribute of electrical theory. Invoking electricity enables us to avoid the mechanical blasting, usually required of gravitational and explosive mechanics, that brings inordinate destruction and thermal excess to situations where we seek quantavolutionary change with a maximum of selectivity and minimal mechanical bursting.

Despite their ubiquity, electrical phenomena have been isolated from the rhetoric of causality. When treated, they have been allowed as only secondary or even tertiary effects; instead mechanical and gravitational processes of enormous magnitude are postulated as the forces playing the primary (causal) role. Sometimes magnetism (usually not observed directly) is seen to play an intermediary, or secondary, role in the deduced causal train which leads to the observed effect. But our outlook has changed. Once practically dismissed as inoperative in celestial matters, electricity, together with electrical effects, has increasingly been recognized to play a role in cosmic actions.

In every natural and biological process - creation, accumulation, structure, function, storage, dissipation - electrical theory is at home. The smallest observable or inferable operation of a molecule, and the largest explosion of a nebula, can be referred to the unified language and lawful behaviors of electricity.

PART THREE:

TECHNICAL NOTES

TECHNICAL NOTE A

ON METHOD

Scientific method goes far beyond such tasks as washing test tubes antiseptically or inventing a better particle shield. It is more than a logical or mathematical calculation. On any question of importance, as here in cosmology, it invokes a sociology of science and a philosophy of being and change.

In the famous Piltdown Hoax, a deliberately buried modern brain case and orangutan jaw were exhumed in 1912 and pronounced an exciting discovery in human evolution (see Johanson and Edey, pp. 77-83). Most scientists, led by an authoritative English group, assigned to the discovery an age of half a million years and Piltdown, England, became a sanctuary of anthropology for a long generation.

Harry Morris was a bank clerk and amateur archaeologist. He collected "eoliths", artifacts of the Neolithic period. But his finds were rejected and ridiculed. The hoaxer of Piltdown had cast some eoliths among the relics; suddenly these were received as *paleoliths* and respected as part of the Piltdown assembly. Morris wrote letters accusing Dawson, the discoverer and a likely culprit, of fraud. To no avail. In 1926 Edmonds published a geological map of the area of Piltdown, which placed the gravels of the discovery site in the upper Pleistocene of fifty thousand years ago, one-tenth of the age assigned to the hoax material. This was not noticed until

1937 when Oakley, doing fluorine research on the Piltdown bones, discovered flagrant discrepancies between the supposed parts of the same being. Finally, in the nineteen-fifties, the hoax was exposed (Weiner, p. 19).

For us the most important lesson of this case and similar ones rests not so much in the immorality of the hoax and cover-up, with their prolonged damaging of scientific anthropology, but in the ever-present sociological process, which here demonstrated how authority in science has the same kinds of effects as it does in religion and politics - to turn attention from anomalous facts, to block inquiries, to discriminate against outsiders, and to maintain and boost reputations.

These effects are normal to authority and countervailing to the also normal productive effects of authority in organizing work and maintaining morale. Embedded in the social process, scientific method is fully susceptible to fashion, also. Fashion is a modern guise of authority - there are fashions in religion and politics, too. It impels scientists to seize enthusiastically upon directing hypotheses as truths that justify a monopoly of attention, making work difficult for others concerned with conflicting hypotheses. Recently, a colleague, James Christenson, who had worked with the 1980 Nobel-award-winning Cronin-Fitch experiments in particle physics, reflected that they indicated nature to be biased in favor of running forward in time. For a generation, highly touted theory had worked upon the hypothesis that "time" was neutral to direction, contrary to human mental expectations. He went on to say that the "big bang" theory of the origin of the "expanding" universe should not have been implicated in these varying experiments. "The Royal Academy made a big deal out of the cosmological stuff because it looked like astrophysics. That's purely speculative and involves an unstable proton." Scientific models of time and motion continually change in these years, often with only the slightest evidence, but pretending a great deal of it.

In 1980 an interdisciplinary conference at the Field Museum in Chicago devoted itself to examining what some members called "macroevolution" and we have called in this book and elsewhere "quantavolution". The proceedings were not to be published, but the thrust of the meeting was publicized as denoting the prominence, if not pre-eminence, and even necessity, of geosphere and biosphere changes, of abrupt, large-scale, intensive events. The new stress is interpretable as a veering towards, and a cautious detour around, the barricaded door of scientific catastrophism with an ultimate crashing through the gates of extra-terrestrialism.

In hundreds of cases since 1942, when a coded message was flashed from Chicago that "the Italian Navigator has landed," scientists have uncoordinatedly begun to tap into the paradigm that looks upon nature as quantavolutionary. In all of these cases, we may perceive that a brilliant research technology is at work, a technical methodology operating with a great many electro-chemico-mechanical devices, but also that this technology inherently must depend upon the ability to ask questions and make mental combinations that position the Universe in new ways, whether examining nuclear particles, cell functions, organisms, or gross shapes of the landscape and skyscape. The theory of Solaria Binaria, typical of cosmology, depends for its success upon fashioning an appealing and effective combination of the advanced technical methodology and the guiding questions and scientific imagery of the age.

THE SCIENTIFIC RECEPTION SYSTEM

Like laymen in a court of law, scientists who cross disciplinary boundaries are chagrined to discover that in another scientific jurisdiction their "best" evidence is inadmissible. For reasons similar to those of a court of law, and with consequences that are often acknowledged by the court itself to be dysfunctional as well as functional, evidence must be limited to certain kinds, pre-processed in a certain manner, and presented in a certain way. To all else, the court is determinedly and deliberately blind.

In schools of law, realization of this large fact of the preeminence of legal procedure can be traumatic to the naive beginning student. In schools of science, the same pre-eminence of procedure will often cause the same shock in the student, but is mitigated by the more confident assurance of the teaching authorities that the process is fully rational, not mythological or conventional in any way or form.

The scientific petitioner, assuming that he has a truth which, if properly heard, would be acknowledged, may try to win his case by several strategies. He may fashion his evidence so as to be heard in the court - framing it as a hypothesis, eliminating value-judgments, quantifying its procedures, obtaining expert witnesses, publishing related material in a most reputable journal, and putting himself forward in academic regalia.

Failing to win a subsequent judgment, the scientific petitioner may resort to a court of different jurisdiction, another discipline - history of science, say, rather than astronomy. Or he may appeal to a higher court, the cosmological and philosophic jurisdictions, for instance. If these resources are denied him, or give judgment against him, he may seek to replace the judges (as for example, Franklin Roosevelt did with the U. S.

Supreme Court), or to create a new court (as Courts of Equity were established to give justice in cases unframable for ordinary judicial consideration).

If rebuffed in these attempts, or if his creations fail him, he can go to "the bar of public opinion," where by an adequate display of persuasiveness, power, and intelligent support, he may intimidate or enlighten the judicial institutions, and obtain in one way or another a rehearing or a favorable verdict that is masked as a rehearing.

Finally, in a revolutionary setting, and with the justification that the system is too rigid for reform, he can try to overturn the juridical order and replace it by a new juridical establishment operating under new rules for the admission and hearing of cases and evidence.

Probably most scientists who have had occasion to test the reception system of science, and whom repeated frustration has not reduced to emotional confusion, will recognize this order of possibilities in pursuing the truth as they see it.

They might also acknowledge that in the past half-century the reception or court system has been elaborated ingeniously, if unconsciously, to provide a modicum of success to everyone - so that there are more judges than petitioners, and a court for every conceivable case and procedure. The bureaucratization of science in academia, government and corporations promotes such a development. This tends to trivialize the caseload of all courts, and sends up a miasma of mutual deference to ward off critics. The resulting rigidity tends to create a revolutionary opposition from the start, a point that has evaded most writers seeking to explain the plethora of anti-scientific books and movements. It is not too far-fetched to compare the situation with that in worldwide politics that has produced so much terrorism.

COSMOGONY: A GHOST FIELD?

In the present work, we have directed ourselves to the discipline, or court, of cosmogony. This, we might think, is logical, since the work concerns ultimate causes of the physical and biological world. Unfortunately, however, the field of cosmogony hardly exists. Such is indicated, for example, in the latest (1974) *Encyclopaedia Britannica,* where neither "cosmogony" nor "cosmology" is allowed a place between the substantial essays on "cosmic rays" and "Costa Rica." Further, in a mere several paragraphs of the "Macropaedia" (vol. 3, pp. 174-5) we are led to perceive these subjects as special areas of astronomy (the "big bang" hypothesis, etc.) or of mythology and ancient speculations about the Universe.

Nor are cosmology and cosmogony offered, much less required as subjects of study in universities; exceptions are rare and usually to be found in schools with a religious bias. Writings on cosmogony are likely to run off the pens of elderly astronomers, "born-again" physicists, and uncomfortable priests. A discipline without a method is a risible contradiction in terms, but such happens to be the situation.

Since the court of cosmogony is largely imaginary, we may expect an *ad hoc* panel, drummed up from various professions, to sit in judgment on our work. For their troubles they will find little that can be termed a cosmogonical method. Rather, they will find in one place the methodology of spectroscopy, in another place that of microbiology, then again that of Egyptian mythology, and now, too, that of theology. It is not because we possess any distinction in these, or in other fields, that we treat of them, but because of the broad and general nature of our problems and of our desire to be as denotative and technically correct as we can be.

At the same time, as must benefit topics so large and fundamental, we avail ourselves of the general operational logic that is accessible to every educated person when working upon any subject whatsoever. We regret, as much as every last reader, the paucity and unreliability of data - in astronomy and physics, we hasten to interject, as well as in mythology and the history of science - and that therefore frequent speculation is necessary, although controlled to be sure, up to the final leap. By way of consolation, one of the auxiliary functions of our study may be to bring to our readers a poignant awareness of how speculative indeed is the basis of the sciences that are concerned with our subject matter.

Thereupon one may appreciate why we must concern ourselves with the simplest of logical and psychological operations in a work of the highest scientific pretensions. For example, the important idea that the Greeks and Romans named planets to correspond to the rank order of importance of the gods is realized only after prolonged study. Saturn, as the retired god *(Deus Otiosus)* of a planet, is second only to Jupiter in size. But how could the ancients have known this without telescopes?

And why would Saturn then be made "father" of Jupiter? Jupiter, the largest planet, is king of the gods, wherever his name or a version thereof is employed. Then come the children, Mars, Mercury and Venus, the others (Neptune, Uranus, the asteroids, and Pluto) being invisible. Mercury

(Hermes, Thoth) is more important, earlier and absolutely, than Mars, even though it is smaller in the sky. This we think is significant.

Striking, too, is the widespread ancient insistence that planet Venus, the brightest and most conspicuous starry object to the eye, is an offspring of Jupiter; for its size and brilliance should have identified it as the ruler of the planetary gods. The significantly larger-sized Sun and Moon are part of most religions, but have not received over the past several thousand years the frenzied and obsessive worship of the others. The Earth, of course, as Mother Goddess, closely identified with the human race, related as a being to, but was not placed in, the category of planets. The recency of Venus is suggested; also, one may surmise that the order of the planets and gods has been overlooked because observers, believing Venus to be a primordial planet, would not notice this coincidence. Thus several simple facts can lend their weight to our theory.

Another example occurs from ordinary psychology. Obsessiveness (and compulsiveness associated with it) is a common behavior. In the history of religion (and what is not associated with religion in earlier times?), obsessive-compulsive behavior is the main trunk of the human mind. Furthermore, this obsessiveness pursues a direct line of extraterrestrial concerns, such as we have incorporated into this book and elsewhere (de Grazia, 1981, 1983b, 1983c, 1983d). Yet many scientists and experts, in putting aside their own subjectivities so as to pursue objective, value-free truths, put aside the subjectivities of their patients (the myth-makers and myth-preservers) and discuss the infinitely varied product of the mythic mind as if it were bubbling up randomly and without reference to objective reality.

Human obsessive-compulsive behavior has causes; it differs from the compulsive instinctive reactions of animals; yet it does not come from a mental *tabula rasa*. It is both logically and psychologically proper to descend the trunk of the human mind in search of those causes until one finds at its roots events adequate to have brought about a heavy dedication of mind and culture to them. Insistent rites, pronunciamentos, testimony, and affirmations demand the recognition of these events as the peculiar causes of compulsions. We think it more plausible that man was watching a sky model and emulating it than that, say, a hominid, who mumbled words and killed his kind, should become casually interested in the sky and use celestial imagery to describe his behavior.

THE HUMANIST-SCIENTIST DIVORCE

In the absence of a field with its special jurists, and of a guiding methodology, the often-decried misunderstanding between the sciences and the humanities is sure to come to the fore. There is no barrier to the negatively conditioned response of physicists to the humanities and of the humanists to the claims of physics (1984d).

An historian of science, Livio Stecchini (1978, p. 117) has written appropriately:

> Most readers of science, except for the very top layer, reveal themselves as being naive realists without any knowledge of scientific epistemology. An expression of this is that some of them declared that Velikovsky's earlier activity in neurology and psychiatry disqualifies him from discussing question of cosmology. However, it was just from an interest in neurology and psychiatry that Kant moved in his investigation of the phenomenology of space and time, which is the foundation of non-Euclidian geometry and Einsteinian physics....

Snow, Polanyi, Barzun, Conant and others have taken their turns at deploring the misunderstanding. Curricula are reformed to correct it. Yet in continues unabated.

The negative conditioning separating these large groupings of savants grows out of a tendency, in the first place, to define one's field in terms of one's special interests, these not necessarily constituting the general interests of the field. A common pattern of individual behavior in both groups is to proceed by an ever-narrowing path towards the proof of a special theory; any cracking of the frame of the theory will being a heavy cost of retracing the path and finding another or a broader way. Hence even an extended approach within the field is not to be countenanced. Only under optimal and rare conditions, too, does a modern discipline possess clearly defined goals, consequently, intra-disciplinary frustrations are common, as paths without ends are pursued, whereupon, in a typical response to frustrations, scientists will reproach outside fields for the faults that they dare not denounce in their own fields.

Inasmuch as internal confusion is a rather general state of affairs in a field of knowledge, it is ordinary for scientists, seeking an opinion upon a matter where an outside field intrudes upon their own, to seek out authorities in the intruding field to obtain opinions concerning the intrusion. However, the very fact that they are challenged in their own field by someone in another field suggests that this person is a maverick from the other, and increases the likelihood that, when they approach the authorities in his home field, they will receive an unfavorable account of the maverick. For instance, authorities in mythology regard legends as expressive of a culture and of some historical value; but they exercise the same control over legendary testimony as do their counterparts in geology and astronomy over the evidence of these latter fields. Hence, it is not especially useful to inquire of them concerning events that not only they themselves deem improbable, but also which they themselves have already heard from geological and astronomical authorities to be impossible. So the vicious circle is set up.

This happens even with "depth" psychology. Jung ends with mental archetypes, Freud with the oedipal complex. These are myths, scientific myths to be sure insofar as they are objective in their formulations, which advance evidence, but such myths are as far from reality as the creation myths of the tribes of Borneo, not to mention those of the Bible. Conversely, should archaeologists or mythologist have the temerity to ask astronomers whether the Moon could be young or geologists whether a great land might be inundated, they can be fairly sure of a negative answer.

We stress that on many facts and principles of cosmogony one has to be especially careful of what authority to interrogate. All fields of scientific study employ fictions - abstractions, concepts, metaphors, models, and probabilities. All fields of study have private languages, which, useful as they may be to insiders, tend to persuade outsiders of a grasp of reality that may be quite weak.

With such conditions prevailing in the field of cosmogony, a method is proper whose premises and goals are clear, whose terms are defined, which offers proof from the "best" evidence available, and whose propositions fairly reflect and summate all "good" evidence from whatsoever quarter or, lacking means to formulate all of it, admits the exclusions and justifies them on methodological grounds.

The method may be called a "model" when the integration of hypotheses is such as to enable the behavior of a part to be predicted from the behavior of the whole and vice versa, "missing parts" to be deduced from described parts, and the whole to operate as an intelligible system through time.

In sum, the procedures demanded by scientific method are clear and accessible, but misunderstandings among the sciences are psychologically and materially indulged. In cosmogony, the situation is grave regarding clarity and accessibility of materials, as well as in psychological and material inducements to discord.

PHYSICS AND LEGENDS

Usually "misunderstanding" between "humanists and scientists" is especially heated on current topics such as euthanasia, crime, nuclear disarmament, vulgarization, and the like; yet nowhere is the malice of natural science towards the humanities so readily vented as when legends are taken seriously. At the risk of controversy, we must nevertheless stress some congruencies between natural science and mythology.

Initially we may compare the structures of legend and science. Any topic of legend can be a topic of science, and vice versa. A legend is an observation or a set of them; so is science. Legend states its observations in human language, rich in metaphor, and carries them orally from one generation to the next and, later, in writing; science seeks non-metaphoric, denotative and quantitative language, and records its observations in information storage and retrieval systems. Legend seeks to retain the functions of moral teaching (" should" and "ought" are persuasive, while "must" is a punitive "should"); science seems to limit itself to precise descriptions and observable relations among events. Legends refer to anthropomorphized sources; science to abstracted forces; both refer, overtly and covertly, to paradigms and ideologies.

Legends are trifled with and tampered with in pleasant times when amnesia overlies historical memories and optimistic wishes can be indulged. In disaster, legends become more important and, under heavy pressures, change significantly. Science changes under the guidance of rules of evidence, the raising of unconscious factors to awareness, and the forging of more and more links in causal chains. Also, science changes by responding to heavy political pressure (Grinnell, pp. 131ff).

The motives behind legends are moral teachings (religious control), and the achieving of a tolerable level of amnesia, involving fun, fantasy and aesthetics, all of which are the more obvious forms that sublimation takes. Although these motives occur in science as well, and science itself is a form of sublimation, science is anxious lest they vulgarize, popularize, distort, and divert its work.

We permit ourselves here, by way of illustration, to speculate and generalize upon an as yet undeveloped series of observations: a systematic study of the oldest nursery rhymes will ultimately discover that every one of these "little classics" (" Chicken Licken", "Hey, diddle, diddle", "Sing a Song of Sixpence", "Ring around the Rosie," etc.) is based upon some historical drama or catastrophe. It will help those scientists and humanists who tend to be snobbish, puritanical or majestic about their material and scornful of the concerns of mythologists with "silliness and superstitions" to reflect upon how much of natural science has come out of amusement, as when early electrical science generated advances from shocking kisses (Heilbron, p236). Myth, science and amusement alike play games with trivia, but the grave cosmos is always unconsciously in mind.

Finally, neither in legend nor in science can the observer have escaped wholly the grip of the ambiance of observations: the observer is part of the observations. The various relativity theories, ancient and recent, make much of this fact. All in all, legend-making and science-making are not foreign to each other but have much in common. Each has its own good reasons for refusing marriage while maintaining liaisons.

Recently, some scientists have named a conjunction of electro-gravitational influences causing natural disorders on Earth the "Jupiter Effect" (see Goodsavage, pp. 144-56). They seem to be able, on good evidence, to demonstrate that Jupiter is not isolated, but has certain fearsome transactional capabilities, which may be exercised upon occasion. An astrologer would say that he has known this all along. Most ancients were obsessed with many "Jupiter effects". We say that these astrological fossils go back to real Jupiter effects that were incomparably stronger than the ones occasioning the present excitement. The ancients, seeking to control the effects, sought to control human behavior aimed at propitiating Jupiter, "lest you die". Our contemporaries do the same, suggesting more pragmatic (effective) means of protecting sectors prone to earthquakes and tidal waves (Gribbin and Plagemann, pp. 132-48).

We would say that the legendary sources are cognizant of grave past effects, and had little new evidence and less control over expected effects. The astrologers inherited confused observations of the past, which further confused them, and could prove no new evidence because they were helpless and incompetent. Our contemporaries possess but disbelieve ancient observations, and also some new evidence of recent times that may have practical value and may lead to a systematic review of ancient celestial behavior. Ancient accounts become simply another source of observations.

The Phaeton legend has been recited to young and old alike for thousands of years: Phaeton, son of Sun, incompetently drives his father's chariot too near to and too far from the Earth, causing great fires and frost. The correspondences between this flight and a cometary encounter are so numerous that many scholars are convinced of Phaeton's historicity, that is, that a comet cut a destructive swath across the tottering globe around the middle of the second millennium before Christ. As Kugler showed, material of scientific value is obtainable from the careful analysis of the legendary stuff on Phaeton (and his namesakes in other myths).

There is adequate reason why the ancient "Jupiter effects" such as cosmic thunderbolts, the Phaeton legends, the natural events reported in Exodus, the Cosmic Egg mythology, the phenomenon of the *Deus Otiosus*, and the divergent "non-astronomical" sacred calendars of the Meso-Americans, Egyptians, and others - to mention only several proto-scientific or disguisedly scientific reports - should be given ordinary treatment, in an integrated manner, in histories of science and textbooks of astronomy, earth sciences, paleontology, and human behavior, including anthropology, prehistory and ancient history. It is perhaps obvious, also, that the ancient accounts of quantavolutionary events find all mankind in the same situations, building related cultures, seeing them destroyed, and recreating them. Once scientists decide to reach back to natural events and primordial human cultures with the hypothesis of Solaria Binaria, they will discover a most inspiring ecumenicalism for our most threatening of times.

TECHNICAL NOTE B

ON COSMIC ELECTRICAL CHARGES

In this work we forgo the concept of opposite charges, which has been in general use since Benjamin Franklin established it. Thus, we revert to a position being argued by other early electricians, who saw no need to introduce "plus" and "minus" charges (Heilbron, pp. 431-38, p. 481). The one-charge idea suits our concept that the Universe possesses a net electrical charge and that all star systems can be represented by cavities which are deficient in that charge. Where the word "negative" occurs in this work it means only the electron and does not imply the existence of an opposing or second type of charge.

For a time we, like others before us, considered the solar charge to be of positive sign, because of the gradual acceleration of the proton wind as it moves away from the Sun. However, this same phenomenon can be viewed as a flow of ions towards a surrounding region of negative electrical charge.

Insofar as solar wind electrons have, if any, only trivial anisotropy in their motion and since detected cosmic-ray ions - which Juergens (1972) has described as the spent wind from the most luminous stars - outnumber cosmic-ray electrons by at least two orders of magnitude, it is logical to conclude that within the region of the Sun most electrons are occupied with sustaining the transaction tending to eliminate the solar cavity. These electrons are *not free:* they form a => *transactive matrix* enveloping the Solar System.

Cells, and maybe even whole biological organisms, are surrounded by charged "skins" or "sheaths" (Ency. *Brit.*, 1974, *Macro.*, vol. 3, pp. 1045 ff.) Their interiors are even more charged than their perimeters, which indicates to us that these biological entities are electron collectors. This, we argue, also applies to the operation of the Sun.

Atoms may be considered in the same way. The atom has long been known to be characterized by electric transactions forming both the interatomic linkages (which create molecules of many kinds) and the interatomic coupling, which defines the "electron-shells" of the atom and may even delineate the chemical elements themselves.

The atom is modeled here as a plenum of charge enveloping a nucleus, which we regard as a massive, dense, compact electrical cavity. Like the cell, the atom exposes to the world a negatively charged perimeter. We therefore chose in this work to avoid speaking of negative and positive ions (say, for example, electrons and protons) being produced when an electron is removed from an atom. Rather we speak of electrons and electron-deficient atoms.

This rhetoric then allows us to describe net charges on bodies that are "negative" (as with the Galaxy, the Sun and the cell) without specifying the sign of the charge. When we refer to ions in this work, we always mean electron-rich atom or molecule. It is noteworthy that atoms are almost always detected and measured when their electrons undergo some form of transition that defines the energy levels and reactions of the atoms. Electrons seem to be the monetary currency of the Universe; stars, cells, and atoms transact and transform to obtain them.

It seems to us that the Solar System's development from creative-nova into binary, through the destructive nova which freed the planets and in the subsequent rearrangement and destructive encounters, is also a story of electron exchanges on the grandest of scales.

The elementary principle governing Solar System behavior is that planets act to accumulate electrons from their surroundings, but in reality they are forced, by the Sun and by their orbital motion, into that space where the electron supply is least capable of yielding electrons to them.[116]

---

[116] Here again, as with stars (as noted earlier in Chapter Three), it is apparent that space itself is the primary determinant of behavior. The stars, planets, and other material in the space compete for the

Planets are also constrained by their electric charges to avoid other planets to the maximum extent. In terms of conventional gravitational models this latter behavior has been described as least-attraction interaction; we see it simply as mutual repulsion between bodies of similar charge density.

Further, planets maximize their capture of the locally precious electrons by developing an electrosphere about their solid surfaces. Atmospheric layers, when present, are within the transactive junction between the planet and its electrosphere. The current flow across the lowest 20 kilometers of Earth's atmosphere is evidence of such a junction. At the outer perimeter of the electrosphere, the "magneto-pause" and "shock front" mark the transactive layer through which the Earth attempts to absorb interplanetary electrons and to exclude solar wind ions. The junction is not always successful: cosmic ray ions regularly break into the Earth's domain, as do bursts of energetic ions generated by solar flares. These ions make the Earth's task Sisyphean: it accretes electrons only to be forced also to take in electron-deficient ions that are hungry as well for the electrons.

An examination of the electrospheres present in the Solar System[117] reveals a "shielding" that protects the charged planets, for they are immersed in a flow of plasma, which must remain close to charge-neutrality. In the plasma, the local differences between electron and ion densities is small, as it is in a metallic conductor through which an electric current flows. Hence in some proportional fashion the small quantity of incident electrons from the Galaxy are distributed to all of the bodies within the cavity by way of the nearly "neutral" plasma. But, in the main, electron accumulation is accomplished by the ejection of ions into the interplanetary plasma from the solar and planetary electrospheres.

By launching ions towards the periphery of the cavity, where electrons are still available, the Sun gains galactic electrons; by contributing to the ion flow the planets gain an appropriate number too. Protons are observed flowing into the solar wind from the electrosphere of the Earth and Jupiter. This outward flow perplexes those analysts who assume electrically neutral planetary environments. Yet it need not, for it can be understood as the only effective method of accumulating electrons within an electron-poor

---

contents of space. These contents not only seem to be atoms and electrons but also a spatial infra-charge, which is not normally available to the body in the space, but whose presence governs all transactions which can occur.

[117] Conventional descriptions of the planetary exospheres describe their electrical properties only as adjuncts to their magnetic properties hence they are there called magnetosphere. Here we consider their magnetic properties secondary manifestations of the fundamental electrified state (see Chapter Thirteen).

cavity. The planet "disguises" its charge level by surrounding itself at great distances with an increasing proportion of ions to electrons. In this way, so to speak, the planet can defend itself in a system where the central Sun voraciously devours any available electrons and jettisons ions onto any reachable electron-sink. The planets, like flotsam, deal with the solar jetsam. Thereupon, the view from each planet is through an electrical fog.[118]

The methodological problem posed in describing quantitatively an electrified cosmos is an experimental problem common to all systems where the instrument disturbs the measured systems. The dilemma cannot be resolved simply by recognizing that the instrument and that which is measured are rendered indistinguishable. We can scarcely imagine how to go about measuring the actual complex of charge-levels existing within the planetary spheres. The problem of determining the charge on a cosmic body resembles the long-established problem of determining how we can feel at rest on the Earth whilst hurtling at fantastic speeds on the globe, in orbit, through the Galaxy, and in the Universe.[119] Should electrical charge prove to be at one and the same time the fundamental element in the Universe and unmeasurable, then we may have to hammer one more nail into the coffin of deterministic physics.

For the first time we are confronting processes occurring at the interactive junctions between large bodies. The very size of the transactions permits humans to observe them broadly, and even to fly among them. (On the microbiological cell level the membrane problem is equally important and complex and there is hampered by technical problems of observation.) Still, the definition of perspectives is difficult in the cosmic sphere, and this is in turn the result of confusing the identities of the actors and the sets. Given the electron and electron-deficient atom as the principal actors, and the scenery of electrospheres, plena and sheaths, the cosmic drama can begin to unfold understandably.

---

[118] The screening of the planets from the Sun resembles the "view" that the valence electron has in, say, a sodium atom; it does not "see" the full nuclear charge because it is screened by the shells of the intervening electrons.

[119] The Earth's equatorial velocity due to rotation is 0.46 km/s, in orbit Earth travels 30 km/s, the Sun moves through the Galaxy at 19 km/s and orbits the galactic center at about 275km/s. The galaxy itself may be traversing the universe at speeds near 540 km/s. Only the first two motions are known with confidence.

TECHNICAL NOTE C

ON GRAVITATING ELECTRIFIED BODIES

In this work we conceptualize "gravitational fields" as an effect of electrical forces acting between charged bodies moving within a charged cosmos (Milton, 1980/81): two bodies respond and move to maintain the greatest separation. In the co-planar orbits of today's Solar System this electrical repulsion among the planets is deemed by us to manifest itself in the Titius-Bode law of commensurable planet periods (e. g. five Jupiter orbits in approximately[120] the same time as two Saturn orbits). Until now the "law" has been an unexplainable observation.

In an electric "gravity" system a tangential inertia[121] is coupled to a radial electrical force whose nature depends upon the electrical state of the bodies orbiting. The electric force can vary between strongly repulsive in close encounter to strongly attractive when electrical flow joins the two bodies (see Table 5 and Figure 38). When the bodies are widely separated and relatively insulated, as are the planets now, the electric transaction among them is repulsive, but is opposed by the surrounding cosmic charge

---

[120] The divergence with theory may be attributable, not to "time of accommodation", but to the complex electrical fields in which the charged planets move.

[121] "Inertia" is usually defined as the quantity of motion (momentum) within a body. It also can be considered as a measure of the difficulty in altering a body's motion (accelerating or decelerating it). For an orbiting body the motion is directed tangentially to the orbit while the force which changes the motion is directed radially.

trying to fill the electron-deficient cavity, which is the Solar System; the two repulsions nearly cancel out, leading to the illusion that something called gravity produces a very weak attraction between the Sun and a planet or between a planet and its satellite(s).

The fact that gravitation, the Great Mother Goddess of physics, has never been found sensibly to exist has nurtured a mild scandal in science for three centuries. After manipulating logically the relevant parameters (the separation of planets from the Sun and their motions in orbit) Isaac Newton concluded that the gravitational force acted everywhere in the same way: it was a universal force (Westfall). That his conclusion was erroneous is becoming apparent. New gravity models incorporate the notion that the strength of the gravitational force (relative to, say, the electrical force) weakens with time (Dirac; Jordan; Dicke, 1957, p. 356; Hoyle and Narlikar; Canuto et al., p. 834).

If indeed the relative strength of the gravitational force declines with time, it means that the mechanical units customarily used to describe celestial motions cannot be interchanged freely with the units employed in atomic physics. Also there is evidence that the gravitational constant varies between experiments (Heyl and Chrzanowski, p. 1, pp. 30-1; Long, 1974). The experiments can be interpreted as evidence that the gravitational constant of proportionality is a function of the spatial separation between the masses gravitating and, in some instances, even of the quantity of mass involved in the "attraction."[122] If gravity is dependent upon time and locality, conclusions about the world based upon a universal force ruling over cosmic motions without intrinsic dependency become erroneous.

More specifically, the mass of a body becomes a function of how its mass is established. Its transactions become environmental rather than absolute. For example, if sex, age and occupation explain a person's consumer behavior, but elements of all are inextricably in all, the decision according to sex alone never occurs but always varies as a function of the other two factors. So here masses measured using transactions in the celestial realm need not be conformable with those determined by transactions between atoms.

---

[122] The implication is that very close and very distant satellites may experience significantly different gravitational transactions with their primary; that is, the force need not remain exactly proportional to the inverse square of their distances as the => *Newtonian formulation* would have it. Since G can have somewhat different values for different separations, then the force function becomes more complex than Newton's Law can handle accurately. Another complexity arises if G also changes values as the amount of mass involved is altered. Such a variation would mean that a binary companion or a Jupiter sized mass would not orbit with a force simply proportional to the force keeping an asteroid or a tiny meteoroid in orbit.

Extrapolations between the cosmic and atomic spheres become meaningless. The bizarre quality of conclusions about recently observed cosmic processes has already spawned the question "Do we need a revolution in Astronomy?" (Clube). All of the dilemmas cited by Clube as confronting astronomers can be resolved in a universe where electric forces are conceived to dominate.

For a long time chemists who concern themselves with the mechanics of collisions between atoms (which are admittedly dominated by the forces between electric charges) have agreed that a collision between two atoms can be treated as a sequence of alternating attractive and repulsive actions (see Figure 37). At great distance the atoms mildly repel one another because their perimeters are sacs of negative charge (blurred electrons). Closer together, electron coupling produces the possibility of bonding and the atoms attract, but further inside, beyond the coupling range, the atoms again repel (this time very strongly). So it is with a "gravitational field", which is then really an electrical field.

The behavior of bodies orbiting in electric transaction differs from those experiencing the conceptually simpler, weak, attractive gravitational force caused only by their mass content. The way in which planets move was shown by Kepler to depend upon the magnitude of the semi-major axis of the orbit.[123] Later, when Newton quantified the "gravitational force" into a relation containing the quantity of matter in each body and the separation of the "gravitating" bodies, Kepler's Harmonic Law was modified to allow celestial systems to be massed (see ahead to Technical Note D).

The law relates three variables: the period over which the complete orbit occurs, T_i ; the average separation of the bodies form the Sun, and the total mass of the system of the Sun and the N -1 orbiting planets,

$$\sum_{i=1}^{N} M_i.$$

---

[123] Its average separation from the Sun.

Figure 37
Potential Energy Curve for the Collision of Two Atoms

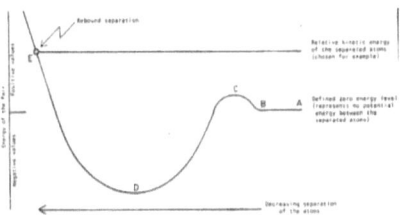

When two atoms collide, electrical force between them acts to alter the energy state of the system compared to the energy which the two atoms posses when they are greatly separated and at rest, the "zero" energy level. Usually two colliding atoms will have more energy than this "zero level"(some positive value). Their kinetic energy of approach determines the closeness the pair can attain in the collision. For a specific energy (the horizontal line drawn above and intersecting with the potential energy curve) the system of two colliding atoms has a surplus of energy represented by the vertical distance between the curves for any chosen distance between the atoms. Where the curves intersect they both represent the same energy; there is no surplus. As the atoms begin to collide, the approaching pair at first do not affect one another (from A to B), but as their electron clouds meet a slight electrical repulsion occurs (from B to C); then electron coupling, as in a chemical bond, produces an increasing attraction between the atoms (from C to D) until a critical separation is attained, when electron decoupling, described elsewhere as internuclear repulsion, begins and produces an increasing repulsion (from D to E) that finally overcomes the inertia (motion) of the pair and causes them to rebound (at E, where the electrical repulsion equals their inertia).

The Harmonic Law, expressed in mathematical terms, states that the square of the period equals the average separation cubed divided by the mass of the system:

$$T_i^2 = \frac{4\pi^2}{G \sum_{i=1}^{N} M_i} a_i^3$$

where the subscript i refers to the motion of the i th planet about the Sun. G is the proportionality factor applying to gravitating systems, and was first evaluated by Henry Cavendish (Shamos).

As traditionally perceived the causal mass terms are invariant hence the other parameters, the separation and period, must as well remain fixed. Given electrically caused orbits, the interbody force depends upon the charge difference on the various bodies in the system. As indicated in our text, we believe that the bodies "gravitate" differently when great charge density differences exist within the system than when they do not (Figure 38).

Figure 38
Electric forces Between Celestial Bodies

(a) Bodies of like charge-density   (b) Bodies of different charge-density

By analogy with the collision between two atoms, charged celestial bodies in collision, if governed by the action of electrical force, also exhibit various possible degrees of attraction and repulsion as they approach one another. In (a) two bodies of like charge and like charge-density experience electrical repulsion as they approach collision. In close encounter polarization of their atoms may redistribute their charges in such a way that some electrical attraction will occur during a part of their approach, but ultimately the two bodies will repel one another and rebound from the collision. In (b) two bodies of like charge but of unlike charge-density initially attract one another as they come together. Polarization may enhance this attraction at closer range and the possibility is great for an electrical discharge between the two bodies as they pass. After the discharge(s) the colliding pair may attain the state of the bodies in (a) and the collision proceeds to closest approach, where the like charges repel the bodies into rebounding apart.

For example, in Solaria Binaria the Sun and Super Uranus never attained electrical equilibrium[124] throughout the lifetime of the binary; their electrical differences persisted, though diminishing with time. The inter-stellar arc was the Sun's attempt to recapture lost charge.[125] It represented an attractive force between the two stars. So long as their electrical natures remained attractive, the inter-star flow continued. If the two had attained equilibrium, that is, had Super Uranus charge-density declined to reach that of the Sun, the two would no longer have attracted one another electrically; their equal charge-densities then would have produced an electrical

---

[124] At equilibrium no net change occurs in a system with the passage of time. Here, interbody electrical currents would cease to flow.

[125] 3 X 10^{22} coulombs might have been exchanged between them over one million years. This represents a transfer of 2 X 10^{44} electrons and a tiny fraction of the mass which flowed between the stars through the plenum. Even with this electrical exchange, the charges moved are negligible compared to the number in a body like the Sun or Super Uranus. If the Sun were an electrically neutral body of mass 2 X 10^{27} tons, the flow would represent an exchange of one electron per one hundred thousand million electrons present. A stellar body carrying net charge, as these were, would be exchanging an even smaller portion of its charge.

"neutrality" in an inertial state.

During the interval when the orbiting stars were seeking electrical equilibrium, the mass of the binary system, as measured using its period of revolution (by Kepler's Law) would have seemingly diminished. As the interval transaction that was accelerating the stars in relation to one another declined, the binary would appear to lose angular momentum. In part this "loss of energy" would be an artifact of the measuring theory; what really was occurring would be a recession of the principals to conserve and gain charge; but a dispersal of charge into the plenum would be occurring as well, causing the plenum to expand and hence the calculated mass of each transacting body to decline.

Taking another example from the Solar System, Jupiter's angular momentum (the product of its mass, distance from the Sun, and its tangential (perpendicular) velocity in orbit) is 2.03×10^{43} (mks units). If it were orbiting at the Earth's distance from the Sun but with this same angular momentum, Jupiter would move at 68 kilometers per second, two and one quarter times faster than the Earth's orbital velocity of 30 km/s. The Jupiter year would be a little longer than 161 Earth-days. The Sun's "mass" required to hold Jupiter, so moving at this closer distance, would have to be five times its present value! If Jupiter were more closely positioned than above, its year would be even shorter, and the Sun's mass would seem even greater.

The story of Solaria Binaria recounts the consequences of the ongoing enhancement of the Sun's charge resulting in the continuously growing repulsion of the planets to regions farther from the solar surface. Analyzed in mechanical terms this repulsion has been reported as a weakened gravitational force over time, it could equally have been seen as a decline in the Sun's mass (its gravitational ability).

Orbits changing under varying electrical transaction behave differently than the conventional view of very slowly evolving gravitational orbital elements. The objects are drawn together or forced apart by changing radial forces. Literally, an object like Venus, born from Jupiter in a charge-deficient condition, spirals inward, driven radially by electrical force and increasing its tangential velocity in sustaining its angular momentum. It is no "lucky billiard shot" that Venus encountered all planets inferior to its initial position near Jupiter. Following an initial diminishing spiral path generally close to the same plane as the other planetary orbits, Venus could not avoid close (i. e., effective) encounter with each body it passed en route to its present orbit. The events described in this book are the recorded, recollected and inferred consequences of many planetary encounters both before and after the excursion of Venus made famous in our time by Immanuel Velikovsky.

TECHNICAL NOTE D

ON BINARY STAR SYSTEMS

In the sample of the sixty nearest stars to the Earth we include the Sun. Accompanying seven of these stars is at least one dark unseen body. These unseen bodies are inferred because a wobble is detected in the peculiar motion of the star associated with the dark body (as in Figure 1). Including the unseen bodies as small stars we find sixty-seven stars grouped into forty-five systems. There are three triples, sixteen doubles, and twenty-six single stars. Sixty-one percent of these objects are thus components in a double or triple star system.

There are potentially many binaries in the Galaxy. Since faint companions are unlikely to be detected by any means, many of the binary systems which exist will not be recognized by observers.

In general, binaries fall into groups separable only by the technique used for their detection. Where the principals sufficiently separate they can be resolved by visual observation through a telescope: these are the visual binaries. When the principals are closer together spectroscopic detection is sometimes possible. For very close pairs eclipses are sometimes seen as the stars orbit one another. In some cases other phenomena are seen which show regular periodicity betraying the binary nature of the system. Discovery of this type has become increasingly frequent in recent years, greatly expanding the number of known binary systems.

Visual observation of the binary companion depends upon several factors: the proximity of the binary system to the Earth; a sufficient separation of the principals to allow resolution of their images by a telescope; and the occurrence of small differences in the luminosities of the principals, otherwise the view of the => *companion* will be obscured by the light of the => *primary*.

Tens of thousands of binary systems can be resolved by telescope into two separate stars. In about twelve percent of these visual binaries the orbital motion can be measured, but only a few satisfactory orbital analyses have been completed.[126] Where the orbit of the companion relative to the primary star can be measured, and where the distance to the principals can be measured, the physical separation of the pair is known. If the period of revolution of the binary is known, then, temporarily accepting Kepler's Harmonic Law, which is based upon Universal Gravitation as the only force binding the principals, the total mass of the binary system can be calculated (Chapter Three). This calculation based upon Kepler's Harmonic Law is the primary clue to the masses of all stars[127] (but see Chapter Two).

Allen (1963) tabulates the distribution of the stars against the calculated total "mass" of the binary system. For systems equal to or greater in mass than the Sun, only thirty-two percent of the stars are not members of double or multiple star systems. In those star systems of lesser "mass" the percentage of single stars rises dramatically.[128] For the mass range 0.5 to 0.25 Sun, eighty-five percent of the stars appear single. No star below 0.1 Sun seems to have a companion *(ibid)*. This surely indicates that the ability to see companions near such poorly luminous stars is limited, if not nil.

For a typical visual binary one revolution of the companion about the primary takes a few decades. The orbits of the companions have dimensions comparable to the orbits of the major planets in the Solar

---

[126] Batten (1967) notes the great difference between the number of systems known to exist and those which have been studied. A highly special sample has well determined orbits, even fewer systems have known masses. Typical orbits are given in Allen.

[127] We use the term massing in preference to weighing. An example of massing using Kepler's Harmonic Law: the satellite Triton is 353 megameters form Neptune; the Moon is 384 megameters from the Earth. If both Earth and Neptune had the same mass, the periods of revolution for Triton and Moon about their primaries would be about the same. They are not; the Moon takes 22.3 days to orbit while Triton orbits in 5.9 days. This leads astronomers to conclude that Neptune is 17 times the mass of Earth. Any transaction equal to 17 times the gravitational pull of *one Earth mass* on Triton would suffice to cause Triton's rapid orbiting of Neptune as observed. So with the stars: the more intensive the transaction between the principals, the more rapidly the pair will orbit about one another.

[128] In systems which show no evidence of any periodic phenomenon, the star's mass has been inferred using theoretical considerations (see Chapter Three).

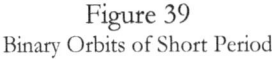

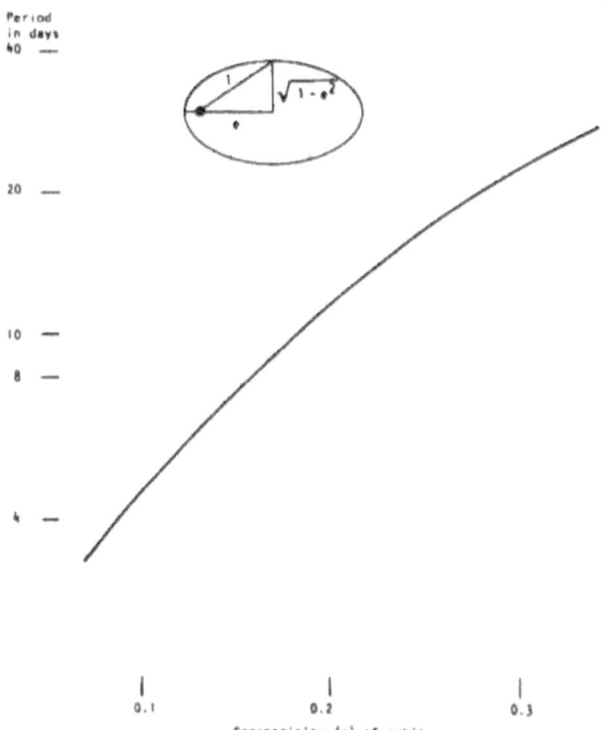

Binary stars show a relationship between the shape of their relative orbit and their period of revolution in that orbit. For those pairs orbiting in times from a few days to a few weeks the orbits are found to be somewhat like the more elliptical planetary orbits found in the solar system. Elliptical orbits are described in terms of their difference from a circular orbit using a quantity called eccentricity. Eccentricities for closed orbits have values between 0 (a circle) and nearly 1 (which would be a parabola). The ellipse above the graph shows how the eccentricity is measured for a particular ellipse.

System, but their shapes are much more elliptical than are the planetary orbits (see Figure 39). For a typical visual binary superposed on the Solar System the => *apastron* (near Neptune) is three times as distant as the => *periastron* (near Saturn).

The shorter the orbital period for revolution, the more circular the orbit of the companion. Systems which revolve in less than ten days have relative orbits whose shape resembles the orbits of the planets Mars and Saturn. Where the orbit is less than 100 days the orbit is less elliptical than the orbit of the planet Mercury. For orbits over 100 days distinctly elliptical orbits are noted and apastron is about twice as distant as periastron. These orbits are more elliptical than the orbit of the planet Pluto, where aphelion is sixty-seven per cent further than perihelion.

In some binary systems the separation of the components is too small to allow resolution in a telescope. Sometimes the detection of the binary still can be made because when the distance between the principals is small enough the stars move in orbit with high velocities. The binary can be observed because a Doppler shift occurs in the spectrum lines of the orbiting companion.

Spectroscopic detection favors binary systems in which the stars are highly luminous and especially where the orbiting star is equal in brightness to, or brighter than, the more stationary primary. The orbital periods for spectroscopically detected binaries range from days to weeks. In such systems the orbital period is determined from the time taken for the spectrum lines to shift through one complete cycle; canceling the motion of the binary system itself, the spectrum of the companion shows a velocity of approach, then no velocity, a velocity of recession, no velocity, finally returning to a velocity of approach.

Nineteen percent of all bright stars show variable Doppler shift in their spectrum, implying a companion (usually unseen). Of these, forty-seven percent show double spectrum lines; the duplication arises because the motion of both of the principals is detected, indicating that the two stars are comparable in brightness.

Lastly, some binary systems are detected because the light received from the stars is seen to vary as the principals eclipse one another. The stars in these eclipsing binary systems usually revolve about one another in less than one month. If indeed these light variations are eclipses, the principals are very close together or, alternatively, at least one, and sometimes both, of the stars have a very large radius compared to the Sun. Orbits have been calculated for almost 100 eclipsing binaries.

About nine percent of the spectroscopic binaries are also eclipsing binaries. To have such a high percentage of eclipsing systems in the spectroscopic binary sample is surely an anomaly.

Eclipsing binaries include principals with the smallest separation; the close binary stars belong to this group. About sixty percent of eclipsing systems can be described as detached, which means that the light curves of the eclipse produced as one star obscures the other show that the principal bodies are roughly spherical in shape; the Algol star system falls into this group of eclipsing star systems.

The remaining eclipsing binaries are the semi-detached star systems. Here the surface of at least one of the principals is distorted into an ellipsoidal shape, and forms at the extreme a teardrop-shaped body "in contact" with the other star. The Beta Lyrae system is a semi-detached binary. Though there is no physical distinction between all of the detached binary systems, that group transacts differently and less strongly than the remainder of the sample, all close binaries. These binary stars transact much more strongly because of the proximity of the two stars. The behavior of the close binaries can be characterized by its violence, in some examples episodic, in others sustained. Here the stars are in competition with the locally available energy supply and for the space with its infra-charge.

Of special interest are the so-called contact binaries, systems in which one of the stars has seemingly expanded so as to touch, or in some cases even to envelop the companion star within its tenuous atmosphere. Some contact binary systems appear to revolve about one another in a small fraction of one day.

Seldom do the close binaries resolve into two stars, nor do their spectra often show duplication. They are the binary systems with the greatest internal transaction. Many of them show gas flowing between the stars (Chapter Ten), some exhibit emission lines, in others one of the components, usually a dwarf star, erupts regularly *(ibid)*. This eruptive behavior seems to be linked to the gas flow, which produces a hot spot on the recipient star, representing a cataclysmic extreme in activity of the type exhibited by the close binary group as a whole.

Systems containing the dwarf novae fall into a group which also resembles systems containing old novae and W Ursae Majoris binaries (Glasby, p. 146). All of the principals are underluminous. In contrast, many close binaries contain one "overly large" principal. The Wolf-Rayet stars are found paired with a smaller overluminous companion (Glasby, p. 143). Frequently, B-emission stars are members of close binary systems (Maraschi *et al.*). As early as 1938 Haffner and Heckmann proposed that in open star clusters, stars lying above the Main Sequence (overluminous stars) were members of binary systems. It seems that a property common to close binary systems is deviant luminosity of one or both principals. This may indicate the importance both of the transaction between the components in such systems, and of the competition of these stars for the contents of their surroundings. We maintain that these transactions are electrical.

In summary, the close binary stars feature one principal which is a degenerate object. At least one of the principals shows anomalous luminosity. Transactions within these systems produce various degrees of violent outburst: some flicker (Chapter Ten), all exchange material and, we believe, electric charge. These unusual characteristics of close binary systems appear to represent a competition for space and electrical charge; some scholars, perplexed by these same behaviors, have proposed that unimaginable concentrations of matter have been observed and are causing the observed violence. From the evidence presented in this book, it seems that Solaria Binaria quantavoluted through the gambit of close binary phenomena before its principals became detached and its binary nature became disguised. The electrified star system, simple in concept and understandable in its development, was the stage on which the pageant of mythology, pre-history, and written history begins to unfold as parts of the common cosmic voyage.

TECHNICAL NOTE E

SOLARIA BINARIA IN RELATION TO *CHAOS AND CREATION*

In 1981, one of us (Alfred de Grazia) published *Chaos and Creation,* which presented the model of Solaria Binaria as part of a general theory of quantavolution. During the last years of its writing, he discussed first with Ralph Juergens and then with Earl Milton the idea of a book on the subject, that would establish it upon firmer foundations and raise it to a new conceptual plane. Juergens' direct participation had hardly begun when he died; but his encouragement and his writings were inspiring to both of us and so we dedicate this book to him in gratitude and friendship.

While *Chaos and Creation* was going through the toils of publication, its author was well aware that he had only spoken the first words on the topic of Solaria Binaria, and that his books would need amendment as soon as a new book could be written. This is not an unusual phenomenon in rapidly developing areas of theory and research, and he is even pleased to constitute a case in point for the pragmatic view that science is never a final statement of truth, and to acknowledge the technical and theoretical superiority of the present work in regard to the model of Solaria Binaria.

Most prominently, our collaboration in the preparation and writing of this book has led to a purely electric theory of the Universe. *Chaos and*

Creation still speaks of electro-gravitational forces, although it relegates gravitation largely to inertial phenomena and stresses the universal electrical energies that are generated and employed in cosmic encounters.

The electrical theory of *Solaria Binaria* further dispenses with two-sign charges, designating only the electron as the independent variable of electricity, and describing relevant natural events by the extent to which they are electron-deficient or electron-rich. We rely exclusively upon electrical charges to motivate transactions within the cosmic realm. We present our propositions, principles and evidence without resort to the concept of gravitation. This is the first work to present a history and dynamics of the Solar System in an entirely electrical form. It offers the first electrical cosmogony.

Looking specifically to *Chaos and Creation*, there de Grazia states how the electrical manifestations declined because, he claimed, the Sun's charge has "always" been diminishing as the galactic input declined. While the galactic transaction was indeed declining and will ever continue to decline (because the Sun's cavity is filling up) the solar charge has increased steadily. But time has evened out the charge distribution within the cavity as well. And so intra-cavity electrical transfers are much less frequent and are of much lower intensity today than ever before. So de Grazia was right in that "electricity" has declined, but not because the solar charge has diminished as he once claimed.

Later, de Grazia described how Super Uranus met its end, using electrically induced rotation to produce mechanical rupture of the star. Here we describe the same process in terms of an electrical instability in Super Uranus' outer layers. Both processes eject debris into the magnetic tube; both would produce sudden fission; but electrical instability would be more easily produced and could focus its effect towards the Sun and the other planets, giving both the recession of the old star and injection of the new partner into the binary position in line with the ancient string of planetary beads lying along the electrical axis.

Again de Grazia talks about differences between electrical and gravitational systems. There, he notes that electrical differences are quickly erased (non-conservative behavior) while gravitational properties exist. In the strictest formal sense, as used in Physics, both fields (electrical and gravitational) are conservative. The strong electrical field in an excited state can relax itself quickly (by emitting electromagnetic radiation as in the atom) while the weak gravitational field cannot. Translated into phenomena, the overt electrical properties of the system would be the first to disappear, supporting the illusion of a non-conservative electrical presence as claimed in *Chaos and Creation*.

Finally, in *Chaos and Creation*, after the explosive extraction of the Moon's material from the Earth's surface layers, the Moon was thought to orbit the Earth, its phases inciting the early humans to a period of lunar worship (circa 11 500 to 8 000 years ago). To conclude that the Moon immediately orbited about the nearby Earth (its motion being somewhat disturbed by the Sun's gravity as it is today) is necessary when the driving force for the orbit arises mechanically or by some mechanical-electrical mix. But in the purely electrical field that we employ here, the Moon can remain suspended in the Earth's sky as we propose. The question of why humans worshipped the early Moon does not depend upon the Moon's motion in that era: its size, its prominence, and its observed birth and subsequent assembly before man's eyes provide sufficient motivation for worship.

The time span of *Solaria Binaria*, unlike that of *Chaos and Creation*, includes the whole of the geological, atmospheric and biological development of the Solar System. The authors feel that, although they may have drawn liberally upon *Chaos and Creation*, they have introduced so many novel concepts and solved so many hitherto unrecognized cosmological problems in the present writing, that this book appears as a complete and independent treatise on cosmogony, which, whether or not *Chaos and Creation* is well known to the reader, can be comprehended in its entirety, from beginning to end. In addition, we have introduced a number of formal, stylistic, structural, and mathematical innovations that make the present book, despite the passage of only several years, the work of a new generation in the theory of quantavolution.

OMNINDEX

ABBREVIATIONS USED IN OMNINDEX

abr.	abridgement
Ap.	appendix
art.	article
bk.	book
cf.	compare
Ch.	chapter
col.	column
ed(s)	Edition(s), editors
Eng.	English
esp.	especially
et al.	and others
f., ff.	and the following pages(s)
Fig.	figure
fn.	footnote
l. /	line
loc. cit	in the same place
o.	omnindex
orig.	originally
partic.	particularly
pl.	plate
Pt.	part
priv.	privately
publ.	published
q. v.	see
repr.	reprinted
rev.	revised
[sic]	thus, indicating an irregularity in this items
sci.	science, scientific
Sp.	Spring
Su.	Summer
tr.	translated, translator
unpubl.	unpublished
v.,vols.	Volume(s)

CERTAIN SOURCES AND THEIR ABBREVIATIONS

Acad. Sci., compt. Rend.	*Académie des sciences, comptes rendus*
Am. Beh. Sci.	*American Behavioral Scientist*
A.Chem. Soc., J.	*American Chemical Society, Journal*
An. Rev. As. Ap.	*Annual Review of Astronomy and Astrophysics*
As. Soc. Pac., Publ. (Proc.)	*Astronomical Society of the Pacific, Publications, (Proceedings)*
As. & Ap.	*Astronomy and Astrophysics*
As. J.	*Astronomical Journal*
Ap. J.	*Astrophysical Journal*
Brit. As. Assn., J.	*British Astronomical Association, Journal*
Can. J. Pl. Sci.	*Canadian Journal of Plant Science*
Chem. & Eng. News	*Chemical and Engineering News*
Creation Res. Q.	*Creation Research Quarterly*
Dept. En. Mines & Res.	*Department of Energy, Mines and Resources (Canada)*
Detroit Ac. Nat. Sci.	*Detroit Academy of Natural Sciences*
Ency. Brit., (Macro. (Micro.)	*Encyclopaedia Britannica*, Macropaedia (Micropaedia)
Ins. El. Eng., J.	*Institute of Electrical Engineers, Journal* (now *Electronics and Power*)
Int. As. U., Proc. 11th Gen. Ass.	*International Astronomical Union, Proceedings of the 11th General Assembly*
Int. Ass. U., Proc. Colloq. N°6	*International astronomical Union, Proceedings Of Colloquium Number Six*
J. Geomag. & Geoel.	*Journal of Geomagnetism and Geoelectricty*
J. Geoph. Res.	*Journal of Geophysical Research*
J. Opt. Soc. Am.	*Journal of Optical Society of America*
J. Phys.	*Journal of Physics*
Nat. Bur. Std., J. Res.	*National Bureau of Standards, Journal of Research*
NY Acad. Sci., Annals	*New York Academy of Sciences, Annals*
Phil. Mag.	*Philosophical Magazine*
Roy. Anthrop. Inst., J.	*Royal Anthropological Institute, Journal*
Roy. As. Soc., Mon. Not.	*Royal Astronomical Society* (London), *Monthly Notices*
Roy. As. Soc., Phil. Trans.	*Royal Astronomical Society* (London), *Philosophical Transactions*
Roy. Soc. (London), Proc.	*Royal Society of London, Proceedings*
Roy. Proc. (New South Wales), J. & Proc.	*Royal Society of New South Wales, Journal & Proceedings*
S.I.S. Review (Workshop)	*Society for Interdisciplinary Studies, Review (Workshop)*

CITATIONS

Abdul-Rauf, Muhammad, 121
(1978), "Pilgrimage to Mecca", *National Geographic* 154, no. 5 (Nov.), pp. 581-607; 113 (pp. 584 f.) also, "Arabia", *Ency. Brit.* 2, 9th ed... p. 262

Abell, George D., 28, 29, 252
(1975), *Exploration of the Universe,* 3rd ed. (Holt, Rinehart &Winston: New York); [1] p. 527, [2] p. 531 f.); => *granule* (p. 256)

Adams, J. A. S., see Heymann

aeon, 36-37, 98-99, 130, 133, 195
is usually an indefinitely long time, here to designate the order of the conventional age of the planetary system, a billion (or thousand million) years. Also equivalent to gigayear.

afterglow, 78, 111
in a molecular gas, is produced by a pulsed electric discharge through pure nitrogen. The afterglow has been observed to persist to the darkness-adapted eye for several hours (Strutt); it is strongly visible for minutes (Ruark et al.). Other common gases produce weaker, shorter-lived afterglows.

Age of Jovea, **xv.** 172ff; 79, 130
is the period following the Deluge (about 5700 BP) to the time of Mercury's encounter with the Earth circa 4400 years ago.

Age of Saturn, **xiv.** 159ff; 152, 162, 165-6, 172
brackets the period eight thousand to fifty-eight hundred years before present.

Age of Urania, **iv.** 50ff; x. 108ff; 11, 57, 60, 75, 142, 144, 146, 150
is the first age of the Quantavolutionary Period, assigned to run from 14,000 to 11,000 years ago. Also called the Uranian age.

Ager, Derek V., 130, 136
(1973), *The Nature of the Stratigraphical Record* (Wiley: New York); (Ch.3; Ch.4)

Ages (Periods of Solaria), 114, 131, 160

Aggarwal, H. R., see Oberbeck

albedo, 146-7
is the fraction of light reflected from a cosmic body.

Agrawal, P. C., *see* Matsuoka

Alfvén Hannes, 76

(1962), "Filamentary Currents and Magnetic Conditions on the Sun" in *Int. As. U.,* Proc. 11 th Gen. Ass. (Academic: New York), pp. 433-5; (pp. 433, 435)

Allen, C. W. Q., 42, 224
  1955, 1st ed.; 1963, 2nd ed.; 1973, 3rd ed.), *Astrophysical Quantities* (Athlone: London); (1963: p. 237); (1963, 1973: no. 112); (1963: 219, 219)

Aller, Lawrence H., 31, 96, 116
  (1974), "Star" in *Ency. Brit.,* Macro. 17 (Chicago), pp. 584-604; (p. 603) see Ross

Alvarez, Luis W., *et al.,* 134
  (1980), "Extraterrestrial Cause for the Cretaceous-Tertiary Extinction", *Science* 208 (6 Jun.), pp. 1095-1108;
  also, Russell, Dale A., "The Mass Extinctions of the Late Mesozoic", *Scientific American* 246 (Jan. 1982), pp. 58-65

Alvarez, Walter, see Alvarez, L. W.

Am. Herculis, 27, 61

anode, 96, 190-2
  is an electron-deficient region in an electric discharge. It is the place towards which electron flow occurs, and can be the source of an ion (q. v.) current - the ions being electron-deficient atoms.

Anaxagoras, 60 fn. 31

apastron, 225-6
  means the greatest separation of the principals (q. v.) in a binary. It is a homologue of apogee for an Earth satellite, and aphelion for a planet. The term apocentron is used elsewhere in place of apastron to describe the farthest point on an orbit.

Apollo, 18, 27, 59, 120, 167, 176-8

arc-second, 59
  is the smallest unit of angular measurement using the scale where the circle is divided into 360 degrees. The degree has 60 arc-minutes. Each minute consists of 60 arc-seconds.

Ares, *see* Mars.

Aristotle, 60, 190
  On the Heavens, tr. W. K. C. Guthrie (Harvard: Cambridge, 1971); (p. 25. 270b) - *Metaphysics,* tr. H. Tredennik (Harvard: Cambridge, 1975); (partic. Bk. One, pp. 3-83) - *Meteorologica,* tr. H. D. P. Lee, (Harvard: Cambridge, 1978 p. 13, 339b21)

Armstrong, T. P., see Krimigis

Arnold, James R., 157
(1973, "The Chemist's Moon," *Science and Public Affairs* 29, no. 9 (Nov,), pp. 22-5

Arp, Halton C., 53
(1975), "The Evolution of Galaxies" in *New Frontiers In Astronomy,* (Freeman: San Francisco), pp. 210-21

Arvidson, R. E., *see* Oberbeck

Asakawa, Y., 90
(1976), "Promotion and Retardation of Heat Transfer by Electric Fields," *Nature* 261 (20 May), pp. 220-1

Asaro, Frank, *see* Alvarez

Asimov, Isaac, 179

Assyrians, 188-9

astronomical unit, 23, 41, 43, 59
is the present value of the Earth-Sun distance. It is equal to 149.6 gigametres (149.6 million kilometres).

Atkinson, G., *see* Jacobs

atmosphere,
Earth's, 24, 31, 61-2, 98, 114, 117, 123-126, 131, 141, 151, 153, 166, 173, 187, 214
solar, 26, 57
also *see* plenum

atomic structure, 28

Axford, W.I., see Krimigis

axis, electrical, **vi.** 67ff; 64, 75, 83, 174-5, 195, 230

axial tilt, 185, 174-6, 132, 89, 86

Ayala, Francisco J., 104
(1978), "The Mechanism of Evolution" *Scientific American,* 23 9 (Sep.), pp. 56-9

Babcock, H. W., 30
(1962), "The Solar Magnetic Cycle" in *Int. As. U., Proc. 11th Gen. Ass.* (Academic: New York), pp. 419-25

Bachmang, Charles H., see Friedman
Bacon, Edward, see Galanopoulos

Bailey, Valentine A., 26, 33
   (1960), "Existence of Net Electrical Charges on Stars," *Nature,* (186 (14 May), pp. 508-10; S; 16, fn. 16 also, *Nature* 189 (7 Jan. 1961), pp. 43-4, 44-5; and *Roy. Soc. (New South Wales) J. & Proc.* 94 (1960), pp-77-86

Baker, George, 126
   (1960), "Origin of Tektites" *Nature* 185 (30 Jan.), pp. 291-4; 120 (p. 293)

Baker, Howard B., 150, 155, 157-8
   (1954), "The Earth Participates in the Evolution of the Solar System", *Detroit Acad. Nat. Sci.* (Detroit).

Baker, Robert H., 38, 40, 126
   (1967), *Astronomy,* 8 th ed. (van Nostrand: Princeton).

barads, 129
   is a biblical term which can be interpreted as the fall of meteorites from the heavens. The Seventh Plague of Egypt. Stones such as are found in great fields on the Arabian desert. See Sieff.

Barnes, A. *see* Wolfe

Barnes, Thomas G., 89-91
   (1977), "Recent Origin and Decay of the Earth's Magnetic Field," *S. I. S. Review* II, no. 2 (Dec.)
   See also, Milsom

Barnwell, F. H. & Brown, Jr., F. A., 98
   (1964), "Responses of Planarians and Snails," in *Biological Effects of Magnetic Fields,* v. 1 (Plenum: New York)

Basilevsky, A. T., see Florensky

Bass, Robert W., 27
   (1974), "Proofs of the Stability of the Solar System," *Pensée* 4, no. 3 (Su.), pp. 21-6; also, loc. cit., "Did Worlds Collide?" pp. 9-20

Batten, Alan H., 27, 54, 57, 62, 64, 77, 112, 114, 224
   (1967), "On the Interpretation of the Statistics of Double Stars," *An. Rev. As. Ap.* 5, pp. 25-44;
   (1973a), "Discussion of Observations of the Flow of Matter Within Binary Systems" in *Extended Atmospheres and Circumstellar Matter in Spectroscopic Binary Systems, Int. As. U. Symposium* no. 51 (May), pp. 1-21;
   (1973b), *Binary and Multiple Systems of Stars* (Pergamon: New York – Toronto)

Beals, C. S. & Halliday, Ian, 121

(1967), "Terrestrial Meteorite Craters and their Lunar Counterparts," in *Int. Dict. Geogr.*, v. 2 (Pergamon: New York)

Becker, Robert O., see Friedman

Becvar, Antonin, 46
(1964) *Atlas of the Heavens 11: Catalogue 1950.0*, (Czechoslovakian Academy of Sciences/ Sky Publishing Corp.: Praha/Cambridge MS)

Beebe, Rita, *see* Smith, B. A.

Behannon. K. W. *see* Ness

Bellamy, Hans Schindler, 126, 157
(1936), *Moon, Myths and Man* (Faber & Faber: London)
(1951), *A Life History of our Earth* (Faber & Faber: London)

Bessell, M. S., *see* Wickramasinghe, D. T.

Bible, references are to Jerusalem Bible (Doubleday: Garden City)

Bimson, John J., 187
(1978), "Redating the Exodus and Conquest," J. St. Old Test., Sup. Ser., no. 5
binary (star systems), 22-4, 41-2, 48-9, 58-60, 71, 77, 112, 195, 212, 223-4, 227-8

biosphere, **xii.** 130ff; 24, 36, 68, 94-5, 99-100, 105, 114, 128, 150, 153, 160, 163, 171, 186, 202

Blavatsky, H. P., 144, 172
(1877), repr. (Theosophical University Press: Pasadena, Calif., 1976) vol. I., Science, pp. 160 f., citing Mallet's *Northern Antiquities*

Blevin, H. A., 63, 70
(1964a), "Plasma in a Magnetic Field" in *Discharge and Plasma Physics,* ed. S. C. Hayden (The University of New England: Armidale, NSW), pp. 471-80
(1964b), "Plasma Confinement" in *op. cit.* pp. 471-80

Bloch, R., 182, 188
(1962), *Gli Etruschi* 11 (Saggitore: Milan)

Blumer, M. & Youngblood, W. W., 186
(1975), "Polycyclic Aromatic Hydrocarbons in Soils and Recent Sediments," *Science* 188 (4 Apr.), p. 53

Bostrom, C. O., see Krimigis

Boyce, Joseph, *see* Smith, B. A.

Brace, Larry H., *see* Taylor

Brady, Joseph L., 23
(1972), "The Effect of a Trans-Plutonian Planet on Halley's Comet," *As. Soc. Pac., Proc.* 84 (Apr.), pp. 314-22

Brahma, 73, 143, 160

brain, 201, 137-40

Brazilevski, A. T., *see* Ksanfomaliti

Brennan, M. H., 64
(1964), "Plasma Heating," in *Discharge and Plasma Physics,* ed. S. C. Hayden (The University of New England: Armidale, NSW), pp. 481-7

Briggs, Geofrrey, *see* Smith, B. A.

Broadfoot, A. L., *see* Kumar

Brooks, J., *see* Hoyle (Oct. 1977)

Brough, James, 104
(1958), "Time and Evolution," in *Studies on Fossil Vertebrates* (Athlone: London), pp. 16-38

Brown, F. A. Jr., see Barnwell

Brown, H. Auchincloss, 176
(1967), *Cataclysms oft he Earth* (Twayne: New York)

Brown, W. Norman, 141
(1961), "Mythology of India," in *Mythologies of the Ancient World* (Doubleday Anchor: New York) Browning, Iben, *see* Roosen

Bruce, Charles E.R., 26-8, 33, 60, 67, 71 110
(1944), *A new Approach to Astrophysics and Cosmogony* (Unwil Bros: London)
(1955), Combination Spectra in Long-Period Variable Stars, *Phil. Mag.* 46 (sep.), pp 1123-31
(1958-1964), Cosmic Electric Discharges, letters to *Electronics and Power* (Ins. El. Eng., J.)

Also:

1. "Cosmic Electric Discharges," v. 4 (Dec. 1958)
2. "Spiral Nebulae," v. 6 (sep. 1960)
3. "Galactic Evolution and Cosmological Controversies," v. 7 (Jul. 1961)
4. "The Energy Radiated by Radio Galaxies," v. 7 (Aug. 1961)
5. "Cosmic Plasma Jets and Hyperthermal WindTunnels," v. 8 (Apr. 1962), pp.204-5
6. "Stellar Temperatures," v. 8 (Oct. 1962), p. 459
7. "Radio Galaxies," v. 8 (Dec. 1962). p. 547
8. "The Evershed Effect," v. 9 (1963), p. 118
9. "Two Populations of Galaxies?," v. 9 (Jun. 1963), p. 259
10. "Zanstra's Theory of Planetary Nebulae," v. 9 (Jul. 1963), p. 303
11. "The Radio Haloes of Galaxies," v. 9 (Aug. 1963), pp. 356-7
12. "Discharge Generated Vortices," v. 9 (1963), p. 414
13. "Extremely Strong Cosmic Radio Sources," v. 10 (Feb. 1964), p. 56
14. "Theories of Radio Galaxies," v. 10 (Apr. 1964), pp. 93-4
15. "Galactic Evolution," v. 10, (May 1964), p. 166
16. "Solar Magnetic Fields," v. 10 (Aug. 1964), pp. 279-80
17. "Whiskers in Space?" [16 sic], v. 10 (Oct. 1964), p. 361

(1966a), Lightning, Novae, and Quasars," *Nature* 209 (19 Feb.), p. 798

(1966b), Lightning Currents," *Electronics & Power* 12 (Jun.), p. 200

also, *Nature* 211 (2 Jul. 1966), pp. 62 ff.

Burgess, Eric, 169, 184
(1979), "Venus Questions Answered," *New Scientist* 81 (8 Feb.), pp. 391-3

Canal, Ramon, 71
(1974), "Nucleosynthesis of Lithium in Low-Energy Flares," *Ap. J.* 189 (1 May), pp. 531-3

Canuto, V., *et al.*, 217
(1979), "Varying G," *Roy. As. Soc., Mon. Not.* 188, pp. 829-37

Cardona, Dwardu, 127, 143-4, 161, 163, 166
(1976), "On the Origin of Tektites," *Kronos* 2, no. 1 (Aug.), pp. 38-44

(1977), "The Sun of Night," *Kronos* 3, no. 1 (Aug.), pp. 31-8

(1978a), "Let There Be Light," *Kronos* 3, no. 3 (Sp.), pp. 34-54

(1978b), "The Mystery of the Pleiades," *Kronos* 3, no. 4 (May), pp. 24-44

catasclysm,
is a sudden dense material deluge from the atmosphere altering biosphere and/ or lithosphere.
see, quantavolution

catastrophe,
    is a sudden large-scale, extremely harmful event; the word probably originated from two Greek roots meaning a "falling star" but came to have assigned to it two different roots, meaning "down-turning" and applied to the denouement of a Greek tragedy.
    see, Quantavolution

cathode, 96, 180, 190, 192
    in an electric discharge is the source of electrons for the conduction process. The cathode usually will be the most electron-rich region.

Cavendish, Sir Henry, 219

cavities, electrical, 25-6, 33, 36-7, 50-2, 100, 118, 131, 212-5, 217, 230

cells, organic, 16, 69, 99, 100-3, 213

Celsius (degree), 32 fn 14
    is the unit of temperature using the scale of 100 degrees between the freezing and boiling points of water at one atmosphere, air-pressure. It was formerly called the Centigrade degree. One Celsius degree is 9/5 of the Farenheit degree still used in both the United States and Great Britain in 1982.

Central Fire,
    also, *axis, electrical*

Challinor, R. A., 93
    (1971), "Variations in the Rate of Rotation of the Earth," *Science* 172 (4 Jun.), pp. 1022-5

Chalmers, J.A., 70, 93, 98
    (1967), *Atmospheric Electricity,* 2nd ed. (Pergamon: Oxford)

chaos, 66, 108-10, 143-4, 162, 164

charge (electrical), *see* electric charge

chemistry,
    of Moon, 155
    of planets, **viii.** 82ff, 95-6
    of stars, Sun, 29-30

Chenault, Roy L., see Ruark

Christenson, James, 202

chromosphere, 27-9, 93
   the gases of the solar chromosphere appear to be hotter than the photospheric gases which lie below them. In the chromospheric region temperature rises abruptly by several tens of thousands of degrees Kelvin. Similar temperature increases have been detected across the chromosphere of other stars (Wright, p. 124). This layer of solar atmosphere can be viewed as an electric double layer between the plasmas of the solar photosphere and the corona.

Chrzanowski, Peter, see Heyl

Chukwu-Ike, Muo, *see* Norman

Clark, D. H., *et al.*, 27, 117
   (1975), "Is Cir X-1 a Runaway Binary?," *Nature* 254 (24 Apr.), pp. 674-6
   (1979), "Differential Solar Rotation depends on Solar Activity," Nature 280 (26 Jul.), pp. 299-300

close binaries, see binary

Clube, S. Victor M., 118
   (1978), "Do We Need a Revolution in Astronomy?," *New Scientist* 80 (26 Oct.), pp. 284-6

Coe, Michael D., 176
   (1975), "Native Astronomy in Mesoamerica" in *Archaeastronomy in Pre-Columbian America,* ed. F. Aventi (University of Texas: Austin)

Coe, M. J., *et al.*, 113
   (1975), "Hard X-ray Measurements of Nova AO535+ 26 in Taurus," *Nature* 256 (21 Aug.), pp. 630-1

Cole, K. D., 93
   (1976), "Physical Argument and hypothesis for Sun-Weather relationships," *Nature* 260 (18 March), pp. 229-30

close binaries, see binary

Collard, H., see Wolfe

Collins, Stewart A., see Smith, B.A.

commensurabilities, see mutual repulsion

companion, 22-4, 27, 30-1, 37, 40, 45, 52, 57-9, 69, 79, 115, 155, 217, 223-4, 226-7
   in a binary system is a body which revolves about the major component (q. v. principal) in the system: the orbiter; as the Earth about the much larger Sun. corona, see solar corona

continents, rafting of, 132, 147, 150-3

Cook II, Allan F., see Smith, B. A.

Cook, Melvin Alonzo, 89, 133-4, 151, 158
(1966), *Prehistory and Earth Models* (Parish: London)
(1972), "Rare Gas Absorption on Solids of the Lunar Regolith," *J. Coll. Int. Sci.* 38, no. 1 (Jan.), pp. 12-19

Cooper, Leon N.
(1968), *An Introduction to the Meaning and Structure of Physics* (Harper & Row: New York)

Corliss, William R., 112
(1975), "Moon Lore and Eclipse Superstitions" in *Strange Universe* (Sourcebook Proj.: Glen Arm, MD)

corona, see solar corona

Cosmic Egg, 109-10, 143, 145, 211

cosmic pressure, 50, 63 fn. 28
on the theory that the Universe is pervaded by a continuum of electric charges, the notion arises that where charge-deficient cavities (stars) exist within the Universe a pressure results driving material within the cavity into one or more aggregations (stars, planets, etc.). The materials within these bodies are confined by cosmic pressure.

cosmic rays, 33-5, 96, 104, 168, 179, 204, 214
are highly energetic electron-deficient atoms (mainly protons) which impinge equally upon the Earth from all directions. The average cosmic ray has an energy of 7 GeV. Cosmic ray electrons exist but they are only one hundredth as abundant as the protons (Hillas. pp. 67-9). The sky "shines" as brightly with cosmic rays as it does with starlight (Watson). The most energetic cosmic rays have an energy at least 100 billion times the average. Such cosmic rays are very rare.

cosmosgony, **xiv.** 159ff; **xv.** 172ff; **xvi.** 181ff; 19-20, 24, 27, 36, 73-4, 195, 204-5, 208, 230-1

Cowley, Ann P., *et al.* 27, 52, 116
(1975), "TT Arietis: An evolved, very short period binary," *Ap. J.* 195 (15 Jan.), pp. 413-21
(1977), "The Flickering White Dwarf CD-42 o 14462: A Non-eruptive Close Binary," *Ap. J.* 214 (1 Jun.), pp. 471-7

Cox, Allan, 88-9
(1969), "Geomagnetic Reversals," *Science* 163 (17 Jan.), pp. 237-45

Crampton, David, see Cowley (1975, 1977)

crater, see astrobleme

Crew, Eric W., 28, 173
(1974), "Lightning in Astronomy," *Nature* 252 (13 Dec.), pp. 539-42

Crozier, W. D., 127
(1966), "Nine Years of Continuous Collection of Black Magnetic Spherules from the Atmosphere," *J. Geoph. Res.* 71 (15 Jan.), pp. 603-10

crust, motions of, xiii. 143ff; 83, 97, 125, 193

Crutzen, P.J., *see* Reid

Curie Temperature, 86, 90 fn 58
(after Pierre Curie) is that temperature at which magnetic materials undergo a sharp change in their magnetic properties. Remnant magnetism appears in rock below this temperature and is erased if the rock is heated above it.

Dachille, Frank, 87-8, 125-6, 176
(1963), "Axis Changes on the Earth from Large Meteoritic Collisions," *Nature* 198 (13 Apr.), 176

(1978), "Electromagnetic Effects of Collisions at Meteoritical Velocities: Experimental and Theoretical Results," *Meteorics* 13 (Dec.), pp. 430-3

(1979), "The Electrodynamic Aspect of Impact Cratering," *unpubl.,* presented at International Meteoritic Society Conference, Heidelberg

Dalgetty, L.C., *see* Dawson

Danielson, G. Edward, *see* Smith, B.A.

Danjon, André, 93
(1960), Note: "Sur un changement du régime de la rotation de la Terre survenu au mois de juillet 1959," *Acad. sci., compt. rend.* 250 (22 Feb.), pp. 1399-1402

darkness, 109, 111-2, 115, 171, 178, 185, 236

Darwin, George H., 92, 155, 157
(1879), "On the Precession of a Viscous Spheroid and on the Remote History of the Earth," *Roy. Soc.* (Lond.), Phil. Trans., pp. 447-538

also, *The Tides and Kindred Phenomena in the Solar System,* 1898, repr. (Freeman; San Francisco, 1962)

Davies, Merton E., *see* Smith, B. A.

Dawson, Edward & Dalgetty, L. C., 85, 201
(1967), "The March of the Compass in Canada," *Canadian Surveyor* 21, no. 5 (Dec.), pp. 380-402

Dayhoff, M. O. *et al.*, 96
(1964), "Thermodynamic Equilibria in Pre-Biological Atmospheres," *Science* 146 (11 Dec.), pp. 1461-4

De Grazia, Alfred
(1977) Ancient Knowledge of Jupiter's Bands and Saturn's Rings, *Kronos*, 2, n°3 (Feb.), pp. 65-8
(1978), "Palaetiology of Fear and Memory" in Recollections of a Fallen Sky, eds. E. Milton *et al.*, (Unileth: Lethbridge), pp. 31-44
(1980), "The 1500 BC Catastrophe: Ten Absolutes for Testing," rev. in "Focus," *S. I. S. Review* IV, no. 4 (Sp.), pp. 74 f.
(1981), *Chaos and Creation: An Introduction to Quantavolution in Human and Natural History* (Metron: Princeton)
(1983a), *God's Fire: Moses and the Management of the Exodus* (Metron: Princeton)
(1983b), *Homo Schizo I: Human and Cultural Hologenesis* (Metron: Princeton)
(1983c), *Homo Schizo II: Human Nature and Behavior* (Metron: Princeton)
(1983d), *Divine Succession: A Science of Gods Old and New* (Metron: Princeton)
(1984a), *The Disastrous Love Affair of Moon and Mars* (Metron: Princeton)
(1984b), *Lately Tortured Earth: Exoterrestrial Forces and Quantavolutions in the Earth Sciences* (Metron: Princeton)
(1984c), *The Burning of Troy: Essays and Notes on Quantavolution* (Metron: Princeton)
(1984d), *Cosmic Heretics* (Metron: Princeton)
[(2009), *The Iron Age of Mars,* (Metron: Princeton)
(1978), *et al., The Velikovsky Affair* (University: Hyde Park, NY, 1967), repr. of art. orig. publ. *Am. Beh. Sci.* (Sep. 1963); rev. ed. (Sphere: London)

deities, 59, 65-6, 72, 74, 110, 114-6, 133, 140-1, 143-4, 159-62, 164, 166, 172-5, 177, 180, 189, 194, 197, 205-6, 248

deluges, 128, 131, 145, 151, 153, 160-1, 164-71, 174-6, 236

Demiurge, 65, 108-9, 114, 164
refers to a grand original intelligence who acted to produce the real world, as described in cosmogonies of early peoples and philosophers.

Dence, M.R., *see* Douglas, J.A.V.

de Santillana, Giorgio & Von Dechend, Herta, 143, 163-4
(1969), *Hamlet's Mill: An Essay on Myth and the Frame of Time* (Godine: Boston)

deuteron, 71
   is the nucleus of a heavy hydrogen atom. Fusion of two deuterons is one step in the thermonuclear fusion of hydrogen.

de Vaucouleurs, Georges, *see* Rudeaux

de Vaucouleurs, Gérard, *see* Menzel

Dicke, Robert H., 97, 217
   (1957), "Principle of Equivalence and the Weak Interactions," *Reviews of Modern Physics* 29 (Jul.), pp. 355-62
   (1964), "The Earth and Cosmology," Ap. to *The Theoretical Significance of Experimental Relativity* (Gordon and Breach: New York), pp. 99-121

Dickerson, Richard E., 99
   (1978), "Chemical Evolution and the Origin of Life," *Scientific American* 239 (Sep.), pp. 70-86

Dirac, Paul A. M., 217
   (1937), "The Cosmological Constants," letter to the Editor, *Nature* 139 (20 Feb.), p. 323

discontinuities, 133, 135-6, 153

Dobrovolskis, Anthony R., see Ingersoll

Dodd, J.R., *see* Napier

Dole, Stephen H., 42
   (1970), *Habitable Planets for Man* (Elsevier: New York)

Donnelly, Ignatius, 128
   (1970), *Ragnarok: The Age of Fire and Gravel,* 1883, repr. (University Books: New York)

double layer (electric), 93
   is the juxtaposition of an electric sheath containing an excess of electrons upon an electric sheath which is electron-deficient. Such a double layer is formed whenever two plasmas of differing electric charge densities meet, for example, between the Sun's photosphere and its corona and between the solar wind and the Earth's plasmasphere. The former double layer forms the solar chromosphere, the latter the Earth's magnetosphere and bow wave.

double star,
   is a synonym for binary star

Dougherty, Ralph C., *see* Edwards

Douglas, J. A. V., *et al.,* 125, 158
(1970), "Minerology and Deformation in Some Lunar Samples," *Science* 167 (30 Jan.), pp. 594-7

Douglas, R. J. W., 123-4
(1970), Sci. Ed., *Geology and Economic Minerals of Canada* (Dept. En., Mines, & Res.: Ottawa)

Dreyer, J. L. E., 73, 154
(1953), *A History of Astronomy from Thales to Kepler,* 1905, repr. 2 nd ed. (Dover: New York)

Dugun, Raymond Smith, *see* Russell, H. N.

Dulk, George A., 180
(1965), "Io-related Radio Emission from Jupiter," *Science* 148 (18 Jun.) pp. 1585-9

dwaf star, 45, 58, 62, 105, 115-6, 227

early-type stars, 58
are those which, using conventional star-evolution-theory sequences, must be younger. Herein, using Bruce's scheme, these are the post-nova stars. They are in our system also high transaction stars.

Earth,
density of, 82, 157
expansion of, 150
extraterrestrial damage to, 119, 178, 182, 185-6
position of,
rotation of, 68, 84-5, 88, 93-4, 147, 149, 153, 174-6
shape of,
surface features, 151 *et passim*

Eck, R.V., *see* Dayhoff

Eddington, 37

Eddy, J. A., *et al.,* 117
(1976), "Solar Rotation During the Maunder Minimum," *Solar Physics* 46, pp. 3-14

Edey, Maitland, see Johanson

Edmonds, 201

Edwards, Deborah, *et al.* 102
(1980), "Asymmetric Synthesis in a Confined Vortex: Gravitational Fields Can Cause a Symmetric Synthesis," *Am. Chem. Soc. J.* 102 (2 Jan.), pp. 381-2

electric neutrality, 33, 93
  as used in this work is a local rather than an absolute condition. The existence of a measurable transaction between local bodies (like the Sun and the Earth) indicate there is not neutrality within the locality. If *the galactic neutral* is one too many electrons per million atoms, while in the Solar System there is one too many electrons per ten million atoms, then a current will tend to flow between the Sun and the Galaxy in order to make the Sun *neutral.*

electrical, ii. 25ff;
  charge, 25, 57, 62, 94, 101, 173, 182, 212ff, 230 *et passim*
  current, 53, 57, 67, 75, 92, 110, 178, 183, 220 *et passim*
  discharge, 33, 60, 63, 70-1, 78, 104, 121, 128, 131, 170, 173, 220 *et passim*
  effects, 180, 198
  forces, 30, 36, 70, 80, 96, 102, 216-7, 219-21

electron-deficient atoms (ions), 51-4

electron(s), passim

electrophoresis, 82, 96
  is the motion of particles (of atomic or larger size) under the influence of an electric field. This motion implies that the particles bear an electric charge.

electrosphere, 124, 146, 151, 154, 166-7, 169-70, 173, 184, 192, 214-5

Eliade, Mircea, 73, 109, 114, 141, 162
  (1954), *The Myth of the Eternal Return,* tr. (Princeton Univ.: New York)
  (1967), The Quest (U. Of Chicago Press: Chicago)

Ellenberger, C. Leroy, 185
  reply to Froshufvud,
  see Talbott, George

Elphic, R.C., *see* Russell (1979)

Emerson, B., *see* Clark

Encke, J. F., 190
  (1823), "Fortgesetzte Nachricht über den Pons'schen Kometon," *Sammlung Astronomischer Abhandlungen Beobachtungen und Nachrichten,* pp. 124-40

Engel, A.R., *see* Coe

eon, see aeon

epoch, see time

Epstein, Samuel & Taylor, Jr., Hugh P., 158
(1970), "O-18/O-16, Si-30/ Si-28, D/H, and C-13/C-12 Studies of Lunar Rocks and Minerals," *Science* 167 (30 Jan.), pp. 533-5

Eraker, J.H., *see* Simpson, J.A.

eruption,
of Super Uranus, 115-117, 119, 128-9, 145-8
of Moon, xiii, 132, 154, 170

Etruscans, 133, 188-9

evolution, 20, 32, 58-9, 99, 105, 136-7, 197-8, 201-2

evolutionary model (E), 99

evolved star, 58
is one which does not obey Eddington's Mass-Luminosity law. Stars in close binary systems are usually of this type, indicative in our view of an intensive electric transaction between the principals in such binary systems.

Exodus, 182, 185-6, 211

expansions, of Earth, see Earth

extinctions, 98, 104, 114, 130, 134-7, 198

faculae, 29, 30
are irregularly shaped unusually bright patches above the solar disc generally associated with sun spots. They are active regions in the photosphere and have their equivalent higher in the atmosphere as chromospheric plages and coronal condensations. (Chromospheric calcium plages are sometimes called flocculi.)

Fairbridge, Rhodes W., *see* Rampino

Fan, C.Y., *see* Krimigis

fallout, 125-6, 128

Faul, Henry, 126
(1966), "Tektites are Terrestrial," *Science* 152 (3 Jun.), pp. 1341-5

Feagin, Terry, *see* Ovenden (1974)

fields, electrical, 28, 96, 218, 230-&, *et passim*

fire, destruction by, 152, 165, 172, 185, 189, 210

Firsoff, V. A., 183
   (1980), "On Some Problems of Venus," *Brit. As. Assn.,* J. 89 (1978), pp. 38-46; repr. *Kronos* 5, no. 2 (Jan.), pp. 57-65

Fisher, Rev. Osmond, 155, 157
   (1882), "On the Physical Cause of the Ocean Basins," *Nature* 7 (12 Jan.), pp. 243-4

fission,
   of Earth, xiii. 143; 160, 167, 170, 230
   of star, 22, 26-7, 50, 58, 60,144

Florensky, K. P., *et al.*, 183, 239
   (1977), "Geomorphic Degradations on the Surface of Venus: An Analysis of Venera 9 and Venera 10 Data," *Science* 196 (20 May), pp. 869-71
   see Ksanfomaliti

Foote, Paul D., *see* Ruark

force, electrical, see electrical force

Forshufvud, Ragnar, 183
   (1979), letter "On the Thermal Aspects of Venus" and "More on the Thermal Aspects of Venus," replies: Talbott, George; Jueneman, Fredrick; Milton, Earl; Ellenberger, Leroy; and Greenberg, Lewis; *Kronos* 4. no. 3 (Feb.), pp. 76-8; *Kronos* 5, no. 1 (Oct.), pp. 83-8

fossil assemblages, 133, 136
   are aggregates of fossils uncovered at a single location. They often exhibit ecological unconformity.

fossil record, 104, 130, 134-7, 189

Fox, Sidney W., 97
   (1960), "How Did Life Begin?," *Science* 132 (22 Jul.), pp. 200-8
   and Windsor, Charles R. (1970), "Synthetis of Amino Acids by the Heating of Formaldehyde and Ammonia," *Science* 170 (27 Nov.), pp. 984-5
fracture, of Earth, *see* Earth

Francis, Gordon, 67, 96
   (1956), "The Glow Discharge at Low Pressure" in *Handbuch der Physik* (Band 22 *Gasentlaufen II* Springer Verlag: Berlin), pp. 53-208

Frazer, Sir James George, 162, 169
   (1916), "Ancient Stories of a Great Flood," *Roy. Anthrop. Inst.,* J. 46, pp. 231-83
   (1922), *The Golden Bough* (Cambridge, 1900), abr. (Macmillan: New York)

Freud, Sigmund, 197, 208

Friedman, Howard, *et al.,* 98
(1963), "Geomagnetic Parameters and Psychiatric Hospital Admissions," *Nature* 200 (16 Nov.), pp. 626-8

Freyer, G.E., *see* Heymann

galactic neutral,
see electrical neutrality

Galanopoulos, A. C. & Bacon, Edward, 186
(1969), *Atlantis: The Truth Behind the Legend* (Bobbs-Merrill: Indianapolis)

galaxy, 18, 25-6, 29-31, 33-4, 39-41, 43, 46-7, 51-4, 59, 61, 69, 71, 110-3, 116-7, 145, 195, 213-5, 223

gases, **vi.,** 67ff; *et passim*

genetic realization, **ix.,** 95ff; 104-5, 130, 138

Gershenson, Daniel E. & Greenberg, Daniel A., 60
(1964), *Anaxagoras and the Birth of Physics* (Blaisdell: New York)

giga(metre), 23, 59, 62, 68, 97, 176
The prefix giga is used to designate thousands of millions; called billions in the United States but not in Great Britain where billion refers to one million million (or 10^{12}). One gigametre is one million kilometres.

Giles, James, *see* Rosen

Gilman, P.A., *see* Eddy

Ginzburg, Louis, 155, 158

Glasby, John, 116, 227
(1970), *The Dwarf Novae* (American Elsevier: New York)

Gliese, W., 42, 45-6
(1957), "Katalog der Sterne näher als 20 Parsek für 1950.0," *Astronomisches Rechen-Institut Heidelberg Mitteilungen* Series A, Number 8, 89 pages

Gloeckler, G., *see* Krimigis

god(s), goddess(es), see deities

Gold, Thomas, 179-80

(1965), as reported in the *New York Times* (21 Apr.), ascribed to remarks at one of the Planetary Sessions of the American Geophysical Union meeting. Not mentioned in the published *Transactions* of the 46th Annual meeting
(1980), "Electrical Origin of the Outbursts on Io," *Science* 206 (30 Nov. 1979), pp. 1071-3; repr. *S. I. S. Review* IV, no. 4 (Sp.), pp. 109-111

Golden Age, **xiv.** 159ff; 154, 174

Goldsmith, Donald, 183
(1977), ed. *Scientists Confront Velikovsky* (W. Norton: New York);

Goodavage, Joseph F.,
(1978), *Our Threatened Planet* (Simon and Schuster: New York)

Graf, Otis, *see* Ovenden (1974)

Grant, Michael, 177, 182
(1974), "Roman Religion" in Ency. Brit. Macro 15 (Chicago)

granule, 28, 32
on the solar photosphere about two and one half million granules exist at any moment. The average granule is 1000 kilometres across; it survives from five to ten minutes. Granules are about 100 K hotter than their surroundings. They show a turbulent motion of about 2 kilometres per second, like a bubble in a porridge pot (Abell, p. 526).

Graves, Robert, 154
(1960), "The Castration of Uranus," Ch. 6 in *The Greek Myths,* v. 1, rev. (Penguin: Harmondsworth)

gravitation, 26, 36-7, 70, 178, 190, 198, 210, 214, 216, 218, 221, 224, 230

Gray, George W., 154
(1955), "Life at High Altitudes," *Scientific American* 193, pp. 58-68

Greece, Greeks, 65, 73-4, 81, 109, 114, 120, 143-4, 154-5, 160, 168, 176-7, 182, 188-9, 205

Greenberg, Daniel, *see* Gershenson
Greenberg, Leonard H.,
(1975), "Diffraction and Angular Resolution" in *Physics for Biology and Pre-Med Students* (Saunders: Toronto), 4-7-1

Greenberg, Lewis M., 161, 183, 187
(1973), Compendium: The Papyrus Ipuwer, *Pensée* 3, n°1 (Winter), pp. 36-7

(1979), "Velikovsky and Venus: A Preliminary Report on the Pioneer Probes," *Kronos* 4, no. 4 (Jun.), pp. 1-2 Reply to Forshufvud, 182 Reply to Morrison, 182 & Sizemore, Warner B. (1975), "Saturn and Genesis," *Kronos* 1, no. 3 (Nov.), p. 46
et al., eds. (1977), Velikovsky and Establishment Science (*Kronos:* Glassboro)
et al., eds. (1978), "Scientists Confront Scientists Who Confront Velikovsky" (*Kronos:* Glassboro)

Gribbin, John R., 93, 176, 210
(1976), "Antarctica Leads the Ice Ages," *New Scientist* 69 (25 Mar.), pp. 695-6
and Plagemann, Stephen H. (1973), "Discontinuous Change in Earth's Spin Rate following Great Solar Storm of August 1972," *Nature* 243 (4 May), pp. 26-7
and Plagemann, Stephen H. (1974), *The Jupiter Effect* (Random House; New York)

Grinnell, George, 209
(1978), Catastrophism and Uniformity: A Proble into the Origin of the 1833 Gestalt Shift in Geology" in *Recollections of a Fallen Sky: Velikovsky and Cultural Amnesia,* eds. E. Milton et al. (Unileth: Lethbridge), pp. 129-37

Gunn, Ross, 60
(1932), "On the Evolution of a Rotationally Unstable Star," *Physical Review* 39 (1 Jan.), pp. 130-41 and "On the Origin of the Solar System," *Physical Review* 39 (15 Jan.), pp. 311-9

Gurnett, D. A. *et al.,* 179
(1979), "Auroral Hiss Observed near the Io Plasma Torus," *Nature* 280 (30 Aug.), pp. 767-70

Haffner & Heckman, 227

Halliday, Ian, 121

Hamilton, D.C., *see* Krimigis

Hammond, Allen L., 178
(1974), "Exploring the Solar System (1): An Emerging New Perspective," *Science* 186 (22 Nov.), pp. 720-4

Hanson, Kirby J., 147
(1976), "A New Estimate of Solar Irradiance at the Earth's Surface on Zonal and Global Scales," *J. Geoph. Res.* 81 (20 Aug.), pp. 4435-43

Hapgood, Charles H., 166, 176
(1966), *Maps of the Ancient Sea Kings* (Chilton: Philadelphia)
(1970), *Path of the Pole,* rev. (Chilton: Philadelphia) 174, fn. 102 (verso front cover, Ch. 1)

Harang, Liev, 81

(1951), *The Aurorae* (Chapman and Hall: London) Harrington, Robert S., see Roosen

Harrison, Christopher G., 150, 153
(1966), "Antipodal Location of Continents and Oceans," *Science* 153 (9 Sep.), pp. 1246-8

Harrison, E. R., 23
(1977), "Has the Sun a Companion Star?," *Nature* 270 (24 Nov.), pp. 324-6

Harrow, Benjamin, *see* Mazur

Hartline, Beverly Karplus, 93
(1980), "Three Spacecraft Team Probes the Magnetosphere," *Research News in Science* 207 (1 Feb.), pp. 511-3

Hatch, Ronald, *see* Patten (1973)

Hawkes, Jacquetta, 116

Hayakawa, S., *see* Matsuoka

Haymes, Robert C., 29, 35, 85-6
(1971), *Introduction to Space Science* (Wiley: New York); 12, fn. 11 (p. 277)

Hays, J. D., 98
(1971), *Geological Society of America,* Bulletin 83, pp. 2433-47

heaven, legendary, 65-6, 72, 108-110, 114-5, 117, 131, 140-1, 143-5, 160-6, 173, 176, 182, 187, 191, 196-7

Heckmann, *see* Haffner

Heilbron, J. L., 210, 212
(1979), *Electricity in the 17th and 18th Centuries* (University of California: Berkeley)

Heirtzler, J. R. & Phillips, J. D.,
(1974), "Rock Magnetism," *Ency. Brit.,* Macro 15 (Chicago), pp. 942-7

Hermes, 177, 206

Hertzsprung-Russell (HR) diagram, 38-9
is a two-dimensional field of stars where luminosity (total radiation emitted) is the ordinate (dependent variable) and color (surface temperature) is the abscissa (determinant variable). This diagram is used extensively in astronomy to infer properties of stars whose distance makes direct measurement difficult or impossible. In terms of the HR diagram, evolved stars are either overluminous or

underluminous for their color, that is, they are above or below the main sequence (q. v.) of the stars.

Hesiod, 110, 114, 154, 182
Theogony, tr. & ed. Hugh Evelyn-Whyte (Harvard: Cambridge 1964) 101 (1. pp. 115, 87 ff.); 106; 148 (1. pp. 720 ff., 131)

Hesser, James E., see Cowley (1977)
et al. (1976), "Sigma Orionis E as Mass-Transfer Binary System," *Nature* 262 (8 Jul.), pp. 116-18

Hewish, A., 70
(1975), "Pulsars and High Density Physics," *Science* 188 (13 Jun.), pp. 1079-83

Heyl, Paul R. and Chrzanowski, Peter, 217, 244
(1942), "A New Determination of the Constant of Gravitation," *Nat. Bur. Stds., J. Res.* (Jul.), pp. 1-31

Heymann, D., *et al.*, 158
(1970), "Inert Gases in Lunar Samples," *Science* 167 (30 Jan.), pp. 555-8

Hillas, A. M.,
(1972), *Cosmic Rays* (Pergamon: Oxford)

Hines, Colin O. (1974), "Ionosphere," *Ency. Brit.,* Macro 9 (Chicago), pp. 809-16

Hoffee, *see* Ransom

Holzer, T.E., *see* Reid

Homer, 172, 179, 188, 191
The Iliad of Homer, tr. Richmond Lattimore (University of Chicago: Chicago, 1951)
The Homeric Hymns, tr. Hugh Evelyn-Whyte (Harvard; Cambridge 1964); 181 (" Hymn to Pallas Athene," pp. 453-4)

homo sapiens, **xii.,** 106, 131-1, 135, 137-8, 141, 144

homo schizo, see homo sapiens

Horowitz, Norman H., 105
(1977), "The Search for Life on Mars," *Scientific American* 297, no. 5 (Nov.), pp. 52-61

Hoyle, Sir Fred, and Narliker, J. V., 217

(1971), "On the Nature of Mass," *Nature* 233 (3 Sep.), pp. 41-4
and Wickramasinghe, N. Chandra (1977), "Origin and Nature of Carbonaceous Material in the Galaxy," *Nature* 270 (22-29 Dec.), pp. 701-3
and Wickramasinghe with other authors, "Prebiotic Polymers and Infrared Spectra of Galactic Sources," *Nature* 269 (20 Oct. 1977), pp. 674-6; "Identification of Interstellar Polysaccharides and Related Hydrocarbons," *Nature* 271 (19 Jan. 1978), pp. 229-31
"Does Epidemic Disease Come From Space?," *New Scientist* 76 (17 Nov.), pp. 402-4 86, fn. 66
also, *Life Cloud: The Origin of Life in the Universe* (Harper & Row: New York, 1978)

Hsieh, S.H., *see* Canuto

Hubbard, S., 136
(1927), *The Doheny Scientific Expedition to the Hava Supai Canyon, Northern Arizona, 1925,* cited by Velikovsky *Kronos* 2, n°2, p. 97, ref. 21)

Hughes, David W., 123-7
(1976), "Earth - An Interplanetary Dust Bin," *New Scientist* 72 (8 Jul.), pp. 64-6
(1979), "Meteoroid Fragmentation in the Auroral Zones," *Nature* 280 (16 Aug.), pp. 539-40

Huitzilopochtli, 189

Hulse, Russell A., *see* Spangler

Hunt, Garry, *see* Smith, B. A.
and Burgess, Eric (1979), "Saturn - Lord of the Rings," *New Scientist* 84 (13 Dec.), pp. 864-7; 165 (p. 867)

Hutchings, J. B., 64, 279
(1976), "Massive Binaries - Early Evolutionary Stages" in *Structure and Evolution of Close Binary Systems* (Reidel: Boston), pp. 9-17

ice, ice ages, 128-9, 132, 136, 152, 166, 176

I Ching, 162
tr. James Legge (Bantam: Toronto 1969);

inertia, 118, 123, 125, 198, 216, 219, 221, 230, 266

Ingersol, Andrew P., *see* Smith, B. A.
and Dobrovolskis, Anthony (1978), "Venus' Rotation and Atmospheric Tides," *Nature* 275 (7 Sep.), pp. 37-8 and see *Nature* 277 (11 Jan.), p. 157

insolation, 147, 184
   is the solar energy received at the Earth's surface. Only a fraction of the insolation is absorbed, some of it reflects into space.

instinct, 138-9, 206

internal neutrality, see electric neutrality

Intriligator, D.S., *see* Wolfe

Io, 179-80, 256, 272

ion, 28-30, 33-4, 39, 52-3, 55-7, 63-4, 69, 75, 77-8, 93, 96, 102-3, 145, 182, 184, 198, 212ff
   is here an atom from which one or more electrons typically present has been removed. *see* also, electron-deficient atoms.

ionosphere, 83, 93
   is a layer of ionized atmosphere beginning at an altitude of 56 to 90 kilometers above the Earth's surface. This layer is electrically conductive. Its altitude and density varies over the day. In theory there is no upper limit to the ionosphere, yet detection of its upper layers is accomplished only infrequently.

Ions, Veronica, 40
   (1968), *Egyptian Mythology* (Hamlyn: London)

irradiance, 147
   is the radiant flux incident upon a unit area of a surface. For sunlight it is the number of watts received per square metre of the Earth's surface.

Isaacson, Israel, 188
   (1974), "Applying the Revised Chronology," *Pensée* 4, no. 4 (Fall), pp. 5-20

Isaksen, I. S. A., *see* Reid

Isenberg, Artur, 182
   (1976), "Devi and Venus," *Kronos* 2, no. 1 (Aug.), pp. 89-103

Jacchia, Luigi G., 123
   (1974), "A Meteorite that Missed the Earth," *Sky and Telescope* 48 (Jul.), pp. 4-9

Jacobs, J. A. & Atkinson, G., 32
   (1967), "Planetary Modulation of Geomagnetic Activity" in *Magnetism and theCosmos* (American Elsevier: New York), pp. 402-14

James, Peter, 142
  quoted by Tresman and O'Gheogan

Jastrow, Jr., Morris, 161
  (1910), "Sun and Saturn," (in English), *Revue d'assyriologie et d'archéologie orientale* (Paris) (Sep.), pp. 163-78

Jerusalem Bible, *see* Bible

Johanson, Donald & Edey, Maitland, 201, 250
  (1981) *Lucy: The Beginnings of Humankind* (Simon and Shuster: New York)

Johnson, Torrence V., see Smith, B. A.

Johnston, M. J. S. & Mauk, F. J., 94
  (1972), "Earth Tides and the Triggering of Eruptions from Mt Stromboli, Italy," *Nature* 239 (29 Sep.), pp. 266-7

Jordan, Pascual, 217
  (1938), "Zur empirischen Kosmologie," *Die Naturwissenschaften* 26 (1 Jul.), pp. 417-21
  Also, "Formation of the Stars and Development of the Universe," tr. H. S. Green, *Nature* 164 (15 Oct.), pp. 637-40

Joss, P. C., 71
  (1977), "X-ray Bursts and Neutron-star Thermonuclear Flashes," *Nature* 270 (24 Nov.), pp. 310-4

Jovean Age, see Age of Jovea

Jueneman, Frederic,
  reply to Forshufvud, Morrison,

Juergens, Ralph E., 10, 28-9, 33-4, 54, 67, 72, 87, 89, 121, 147, 179-80, 190-2, 212, 229
  (1972), "Plasma in Interplanetary Space; Reconciling Celestial Mechanics with Velikovskian Catastrophism," Pensée 2, no. 3 (Fall), pp. 6-12 also, in *Velikovsky Reconsidered* (Doubleday; Garden City, 1976), pp. 137-55
  (1974), "Of the Moon and Mars," Part One, *Pensée* 4, no. 4 (Fall), pp. 21-30; Part Two, *Pensée* 4, no. 5 (Winter), pp. 27-39

(1977a, 1977b), "Plasma Probes," Ap. II (pp. 26-9) of "On the Convection of Electric Charge by the Rotating Earth," *Kronos* 2, no. 3 (Feb.), pp. 12-30
(1977c), "Galactic Space Charge and Stellar Energy," *S. I. S. Review* I, no. 4 (Sp.), pp. 26-9
(1977d), "The Critics and Stellar Energy -Juergens Replies," *S. I. S. Review* II. no. 2 (Dec.), pp. 49-51
(1978), "Geogullibility and Geomagnetic Reversals," *Kronos* 3, no. 4 (May), pp. 52-64
(1979a), "Stellar Thermonuclear Energy: A False Trail," *Kronos* 4, No. 4 (Jun.), pp. 16-25
(1979b), "The Photosphere: Is it the Top or the Bottom of the Phenomenon We Call the Sun?," *Kronos* 4, no. 4 (Jun.)
(1980), "On Morrison: Some Final Remarks -Ralph Juergens Replies," *Kronos* 5, n° 2 (Jan.), pp. 68-75
See also de Grazia

Jung, Carl, 208

Jupiter, **xv.** 172ff;
electrosphere, **xv.** 172ff; 146, 214
as god, **xv.** 172ff; 72, 121, 132, 146, 159, 161, 168, 171
planet, **xv.** 172ff; 19-20, 23, 33, 35, 167-8, 182, 184, 195, 205-6, 210-1, 214, 216-7, 221
Red Spot, 182

Kapitza, P. L., 68
(1979), "Plasma and the Controlled Thermonuclear Reaction," *Science* 207 (7 Sep.), pp. 959-64

Kasuara, I., *see* Matsuoka

Keath, E.P., *see* Krimigis

Kelley, Michael C., 93
(1980), book review of *Solar System Plasma Physics* by E. N. Parker et al. (Elsevier: New York, 1979) in *Science* 207 (18 Jan.), pp. 297-8

Kelly, Allan O., 193
(1974), *The Gravitational Description of Mars,* priv. publ. (Varsbad, CA)

Kelvin, 15, 29, 88
is the unit of temperature using the scale zeroed at absolute zero. It is the lowest conceivable temperature. The Kelvin unit is identical to the Celsius degree. The freezing point of water is 273.15 K(elvin).

Kepler, Johannes, 218, 221, 224

Kloosterman, Johan B., 135

(1976), "Why Did the Dinosaurs Perish?," *Komsomolskya Pravda* (5 Apr. 1965), repr. *Catastrophist Geology* 3, no. 1 (June), p. 5 (supplied courtesy of L. I. Salop), 127, fig. 26

Koch, Robert H., 113
(1970), "Observational Facts in Binary Mass Loss," in *Mass Loss and Evolution in Close Binaries*, eds. K. Gydenterne & R. M. West, *Int. As. U., Proc. Colloq.*, no. 6 (Copenhagen University Observatory)

Kofahr, Robert E., 170
(1977), "Could the Flood Waters Have Come From a Canopy of Extraterrestrial Source?," *Creation Res. Q.* 13 (Mar.), pp. 202-6

Kolin, Alexander; 98
(1968), "Magnetic Fields in Biology," *Physics Today* 21 (Nov.), pp. 38-59

Kondratov, Alexander, 121, 186
(1974), *The Riddles of Three Oceans* (Progress: Moscow)

Kopal, Zdenek, 27, 58, 116
(1938), "On the Evolution of Eclipsing Binaries," *Roy. As. Soc., Mon. Not.* 98, no. 8, pp. 651-7
(1959), *Close Binary Systems* (Chapman and Hall: London)

Kraft, Robert P., 58, 114
(1977), "Double Stars" - a book review of *Structure and Evolution of Close Binary Systems*, ed. by Eggleton et al., Science 197 (29 Jul.), pp. 449-50

Krimigis, S. M. et al., 180
(1979), "Low-Energy Charged Particle Environment at Jupiter; A First Look," *Science* 204 (1 Jun.), pp. 998-1003

Kronos, 66, 160, 166

Krzeminski, W., 116
(1965), "The Eclipsing Binary U Geminorum," *Ap. J.* 142, pp. 1051-67

Ksanfomality, Leonid Vasilevich, et al., 183
(1977), "The New Venus," *New Scientist* 73 (20 Jan.), pp. 127-9

Kugler, Franz Xavier, 211
Kukarin, B. & Parenago, P., 116
(1934), "Investigations of Nova-like Variable Stars, II. Cycle-amplitude Relation in U Geminorum Variables," *Veränderliche Sterne* 4, no. 8 (44) (1 May), pp. 251-4

Kukla, G. J. & Matthews, R., 176
(1972), "When Will the Present Interglacial End?," *Science* 178 (13 Oct.), pp.190-1

Kumar, S. & Broadfoot, A. L., 83
(1978), "Evidence from Mariner 10 of Solar Wind Flux Depletion at High Ecliptic Latitudes," *As. & Ap.* 69, Letter, L 5-8

Kundt, Wolfgang, 183
(1977), "Spin and Atmospheric Tides of Venus," *As. & Ap.* 60, pp. 85-91

Kupferman, P.N., *see* Morabito

Kurtt, W.S., *see* Gurnett

Lagrangian point, 80
in a three-body system the orbits can be computed if one of three bodies is negligibly tiny - in such a case the motion of the minuscule third body does not disturb the two primary bodies. Lagrange showed that for such a "restricted system of three bodies" there existed several points, co-rotating with the motion of the primary pair, where the third body could be trapped. The L1 point is one of these points; it lies between the two primary bodies.

Lamers, H. J. G. L. M., *et al.*, 30
(1976), "Stellar Winds and Accretion in Massive X-Ray Binaries," *As. & Ap.* 49, pp. 327-35

Lamport, J.E., *see* Simpson, J.A.

Lane, A.L., *see* Stone

Lane, Frank W., 180
(1968), *The Elements Rage: The Extremes of Natural Violence,* in two vols., rev. (Sphere: London)

Lang, Andrew, 162
(1968), *Custom and Myth,* reprint of 1885 London ed. (AMS: New York)

Lanzerotti, L. J., see Krimigis

Lapointe, P. L., *et al.*, 86, 174
(1978), "What Happened to the High-Latitude Palaeomagnetic Poles," *Nature* 273 (22 Jun.), pp. 655-7

Larson, E. D., see Strangway
Latham, Gary V., 158
(1973), "Lunar Seismology," *Science and Public Affairs* 28, no. 9 (Nov.), pp. 16-21

Lear, John, 179
(1964), "What the Moon Ranger Couldn't See," *Saturday Review* 47 (5 Sep.), pp. 35-43

least interaction action (sometimes least action interaction) see mutual repulsion

Lebo, G. R. *et al.*, 180
(1965), "Jupiter's Decametric Emission Correlated with the Longitudes of the First Three Galilean Satellites," *Science* 148 (25 Jun.), pp. 1724-6

Lemaire, J. & Scherer, M., 33
(1971), "Kinetic Models of the Solar Wind," *J. Geoph. Res.* 76 (1 Nov.), pp. 7479-90; 16, fn. 16 (pp. 7480 f.)

Lepping, R.P., *see* Ness

Levy, E. H., 83
(1978), "Magnetic Field in the Primitive Solar Nebula," *Nature* 276 (30 Nov.), p. 481

Lewis, Jr., Paul D., *see* Milton (1978)

life,
electrical processes, **ix.** 95ff;
origin of, **ix.** 95ff, 30, 60, 67-8, 72-3, 108, 110, 117, 130, 133-4, 137, 154, 166, 174, 182

light, 28-30, 34, 224, 226-7
celestial (primordial), 59, 144-5, 166, 169, 171
in plenum, 61, 62, 64-5, 74, 93, 108, 111
solar, 33, 93, 175, 179

lightning, 10, 27, 37, 68-71, 88, 96, 98, 120, 124, 132, 150, 156, 172-3, 183-4, 193, 242, 246

light-year, 15, 33, 38, 41-3
is a unit of distance. It represents approximately 10 16 metres, the distance light travels (in theory) through a vacuum in one year (3.16 x 10 7 seconds).

Liller, William (1977), 61
"The Story of AM Herculis," *Sky and Telescope* 53, no. 5 (May), pp. 351-4

Lindblad, B. A., 123
(1979), "Meteor Radar Rates, Geomagnetic Activity and Solar Wind Sector Structure," *Nature* 273 (29 Jun.), pp. 732-4

Lippincott, E.R., *see* Dayhoff

local neutral, see electrical neutrality

Locke, John, 154

Lokanadham, B., *see* Matsuoka

Long, C. H.,
(1963), *Alpha: The Myths of Creation* (G. Brazillier: New York)

Long, Daniel R., 217
(1974), "Why do we believe Newtonian gravitation at laboratory dimensions?," *Physical Review,* Part D 9 (15 Feb.), pp. 850-2
also, "Experimental examination of the gravitational inverse square law," *Nature* 260 (1 Apr. 1976), pp. 417-8

Lowery, Malcolm, 9, 144, 182
(1980/81), "What's in a Name?," *S. I. S. Review* V, no. 2, pp. 46-9

Lugmair, G. W., *see* Marti

luminosity, 23, 34, 37-8, 40-1, 45-6, 63-4, 111, 115, 227-8
of a star depends upon the area of the star's surface (opaque radiating layer of gases) and upon the fourth power of its surface temperature. The luminosity of a star is a measure of its energy output, it can be known directly, as opposed to inferred, only if the star's distance can be measured.

Lyttleton, Raymond A., 27, 60
(1936), "The Origin of the Solar System," *Roy. As. Soc., Mon. Not.* 96, pp. 559-68
(1938), "On the Origin of Binary Stars," *Roy. As. Soc., Mon. Not.* 98, pp. 646-50
(1953), *The Stability of Rotating Liquid Masses* (Cambridge University: Cambridge)

Mack, Ruth, *see* Ponnamperuma

magnetic,
field, vii., viii., 53-7, 60, 63-4, 67-8, 70, 97-9, 102-3, 111, 114, 118-9, 123, 125-6, 145, 150, 155-6, 158, 170-1, 174-5, 230
poles, 15, 129, 132, 147, 149, 174-5

magnetism, 30, 56, 79, 83, 86, 88-90, 92, 98, 174, 184, 198, 235, 246

magnetite, 88
is a black to brownish metallic stone with magnetic properties. The legendary lodestone is one of the magnetites. The magnetites are formed of octahedral crystals of mineral whose chemical structure contains the unit, XFe_2O_4. X may be Fe, Mg, Ni, Zn, or Mn. The first is most common; the last two are only weakly magnetic.

main sequence stars, 58
obey Eddington's Mass-Luminosity Law. They constitute the majority of stars whose distance, brightness, and temperature have been measured.

Makino, *see* Matsuoka

Malin, S. R. C. & Srivastava, B. J., 98
(1979), "Correlation Between Heart Attacks and Magnetic Activity," *Nature* 277 (22 Feb.), pp. 646-8

Malkus, W. V. R., 88
(1968), "Precession of the Earth as the Cause of Geomagnetism," *Science* 160 (19 Apr.), pp. 259-64

Manabe, R., *see* Matsuoka

Manson, Lewis A., 151
(1978), *The Birth of the Moon* (Dennis-Landmann: Santa Monica)

Maraschi, L., *et al.*, 58, 227
(1976), "B-Emission Stars and X-ray Sources," *Nature* 259 (29 Jan.), pp. 292-3

Marlborough, J.M., *see* Cowley (1975)

Mars, planet, **xvi.** 181ff; 14, 18, 27, 59, 105, 132, 167, 170-1, 178, 205-6, 226 *et passim*

Marti, K., *et al.*, 30
(1970), "Solar Wind Gases, Cosmic Ray Spallation Products, and the Irradiation History," *Science* 167 (30 Jan.), pp. 548-50

Martin, P. S. & Wright, Jr., H. E., eds., 136
(1967), *Pleistocene Extinctions: The Search for a Cause* (Yale University: New Haven)

Mason, Herbert, tr., 161, 176
(1972), *Gilgamesh* (New American Library: New York)

mass, 23-4, 30-2, 57-9, 77, 96, 100, 113, 118, 130, 134, 136, 141, 150, 167, 217-21, 224

massive ion, 96
ions are divided into fast and slow. Ions with greatest inertia to the field are said to be massive because they are harder to move; the easier they become mobile the more lightness they are assigned. Elements of low atomic number are most mobile.

Masursky, Harold, *see* Smith, B.A.

Matsuoka, M., *et al.*, 64
(1974), "Further Simultaneous Hard X-ray and Optical Observations of Sco X-1," *Nature* 250 (7 Jul.), pp. 38-40; 47

Matthews, R., *see* Kukla

Mauk, F.J., *see* Johnston

Maya(s), Mayan, 176

Mazur, Abraham & Harrow, Benjamin (1971), 102
Textbooks of Biochemistry, 10th ed. (Saunders: Philadelphia); 91-2, fn. 71 (pp. 71 ff., esp. 74-6)

McCauley, John, *see* Smith, B.A.

McKibbin, D., *see* Wolfe

megawatts, 89
the prefix mega indicates a multiplier of one million. Hence a megawatt is one million watts and a megametre is one million metres

memorial generations, 140
is the difference in years between a youngest listening child and the oldest storytellers of a society. Here we assign this interval a value of 50 years.

Menzel, Donald H., 28, 41, 43, 159
(1959), *Our Sun,* rev. ed. (Harvard: Cambridge); 11 (p. 24)
et al., (1970), *Survey of the Universe* (Prentice Hall: Englewood Cliffs); 24, fn. 23 (p. 659)

Mercury, planet, 18, 27, 59, 167, 170-1, 176-81, 183-4, 205, 226

Meservey, R. (1969), "Topological Inconsistency of Continental Drift on the Present-Sized Earth," *Science* 166 (31 Oct.), pp. 609-11

meteorite, 32, 83, 115, 120-1, 123-4, 128

meteoroid, 13-4, 87, 104, 120, 123-5, 217

methodology, technical note A, 203, 205, 207

Michael, Helenn V., *see* Alvarez

Mihalas, Dimitri & Routley, Paul McRae, 41
(1968), *Galactic Astronomy* (Freeman: San Francisco)

Mihalov, J., *see* Wolfe

Mikami, Y., *see* Matsuoka

Miller, Stanley L. & Urey, Harold C., 67, 96-7
(1959), "Organic Compound Synthesis on the Primitive Earth," *Science* 130 (31 Jul.), pp. 245-51

milli(tesla), 91, 98
   the prefix milli refers to the multiplier one-thousandth. One millitesla is thus one-thousandth of a tesla

Milsom, John,
   (1977), "A Commentary on Barnes' Magnetic Decay," *S. I. S. Review* II, no. 2 (Dec.), p. 46

Milton, Earl R. (1978), "Foreword" in *Recollections of a Fallen Sky: Velikovsky and Cultural Amnesia,* eds. E. Milton et al. (Unileth: Lethbridge), pp. 11-18
   (1979), "The Not So Stable Sun," *Kronos* 5, no. 1 (Oct.), pp. 64-78; 11, fn 7 (pp. 70 f.); 56-7, fn 41
   (1980/1), "Electric Stars in a Gravity-less Electrified Cosmos," *S. I. S. Review* V, no. 1 (Jan.), pp. 6-12; reply to Forshufvud; reply to Morrison
   (1982), "Comets, Rings, Satellites, and Things," *unpubl.*

Mitton, Simon, *ed.,* 52, 116
   (1977), *Cambridge Ency. Astronomy* (Prentice Hall: Toronto)

Miyamoto, S., *see* Matsuoka

mobility (of an atom), 96
   is the ratio of the average drift velocity (attained between collisions) to the electric field strength (which produces the drift velocity).

models, **xvii,** 194ff

Mohorovicic discontinuity, 153
   the junction which separates the Earth's crust and mantle. Its depth is about 10 kilometres below the ocean basin. neutrinos, *see* nuclear fusion

Moon, **xiii.** 143ff, 11, 13-4, 18, 59, 66, 73-4, 122, 132-3, 161, 163, 170-1, 177-80, 183, 189-92, 194-6, 206, 208, 231

Moore, Patrick, *see* Wilkins

Morabito, L. A. *et al.,*. 179
   (1979), "Discovery of Currently Active Extraterrestrial Volcanism," Science 204 (1 Jun.), p. 972

Morris, Harry, 201

Morrison, David, 183
   (1980), letter, "On Morrison: Some Final Remarks," replies: Juergens, Ralph; Milton, Earl; Talbott, George, and Ellenberger, Leroy; Jueneman, Fredrick; Greenberg, Lewis, *Kronos* 5, no. 2 (Jan.), pp. 66-92
   See Smith, B.A.

Mullen, William, 160-1
(1973), "A Reading of the Pyramid Text," *Pensée* 3, no. 1 (Winter), pp. 10-16

Muller, Max, 116

Murray, Bruce C., 176
(1975), "Mercury" in *The Solar System* (Freeman: San Francisco), pp. 37-46

multiple star system, see binary star system

mutual repulsion, 60, 214

mythology, 65, 72, 109-10, 112, 114, 141, 145, 156, 159-61, 164-5, 167, 169, 181, 186, 188, 190, 203, 205, 208-11, 228

Nagata, Takesi, 79, 89
(1961), *Rock Magnetism,* rev. (Maruzen: Tokyo)

Napier, W. McD. & Dodd, R. J., 177
(1973), "The Missing Planet," *Nature* 242 (23 Mar.), pp. 250-1

Narlikar, J.V., *see* Hoyle (1971)

Nather, R. Edward & Warner, B., 62, 113
(1969), "DQ Herculis: Synchronous Photometry," *Science* 166 (14 Nov.), 876-7
see Warner

Neptune, 155, 167-8, 182, 205, 224-5
Ness, Norman F., *et al.,*
(1976), "Observations of Mercury's Magnetic Field," *Icarus* 28, pp. 479-88

neutrinos,
see nuclear fusion

Newcombe, Simon, 32, 179
(1878), *Popular Astronomy* (Harper & Brothers: New York)

Newton, Sir Isaac, 217-8

Newtonian formulation, 217
states that the gravitational attraction between two celestial bodies depends upon the product of the two point masses transacting and upon the inverse of the square of the distance separating the masses. Expressed mathematically

$$F_g \propto (M_a)(M_b)/(d_{ab})^2$$

In metre-kilogram-seconds units (mks) the gravitational constant of proportionality (G) relates the force in newtons to the masses in kilograms and the separation in metres. G has the value 6.667 x 10 -11 m 3 /kg-s 2 so,

$$F_2(N) = G\theta M_a(kg)\theta M_b(kg)/d_{ab}(m)^2$$

Ney, Edward P., 127
(1977), "Star Dust," *Science* 195 (11 Feb.), pp. 541-6

Niemann, V. D., 127
Soviet Astronomy 68

Nieto, M. M., 177
(1974), "The Titius-Bode Law and the Evolution of the Solar System," *Pensée* 4, no. 3 (Su.), pp. 5-7

Ninniger, N. H., 121
(1952), *Out of the Sky,* repr. (Dover: New York)

Nishimura, J., *see* Matsuoka

Norman, John, *et al.,* 121, 125
(1977), "Astrons - the Earth's Oldest Scars?," *New Scientist* 73 (24 Mar.), pp. 689-92

Nova(e), Novas, **xiii.** 143ff; **xiv.** 159ff; 18, 27, 50-1, 56, 59, 61-2, 96-7, 99, 102, 108, 113, 115-8, 127-8, 135, 213, 227

nuclear fusion, 32, 59
is the supposed stellar process by which the nucleii of four hydrogen atoms collide with sufficient energy to coalesce forming a single helium nucleus having slightly less mass than the original hydrogen. The mass which is destroyed in fusion reappears as radiant energy which slowly flows away to the surface. In the fusion, two protons are changed into two neutrons, two anti-electrons, and two neutrinos. The neutrons remain in the fused helium nucleus, the anti-electrons annihilate with two electrons (liberating more radiant energy), and the neutrinos escape the star immediately, travelling at the speed of light.
On Earth, a type of nuclear fusion has been sustained for one hundred pico-seconds. No continuing fusion process has been produced. To remain luminous by

conventional theory the star must fuse hydrogen continuously (Rudeaux and de Vaucouleurs, pp. 316-9). nucleosynthesis, *see* nuclear fusion

nucleosynthesis, see nuclear fusion,

nucleotides, 103
the monomeric unit which makes up the nucleic acid molecules. A nucleotide consists of a nitrogen base, plus a sugar, and a phosphate group.

Oakley, Kenneth, 202

Obayashi, Tatsuko, 30
(1975), "Energy Build-Up and Release Mechanisms in Solar and Auroral Flares," *Solar Physics* 40, pp. 217-26

Oberbeck, V. R., *et al.,* 178
(1977), "Comparative Studies of Lunar, Martian, and Mercurian Craters and Plains," *J. Geoph. Res.* 82 (10 Apr.), pp. 1681-97

ocean basin(s), 65, 128, 132, 136, 151, 153, 169

Oda, M., *see* Matsuoka

Ogawara,Y., *see* Matsuoka

O'Gheogan, Brendan, *see* Tresman

Ohtsuka, Y., *see* Matsuoka

O'Keefe, John A., 127, 157
(1973), "After Apollo: Fission Origin of the Moon," *Science and Public Affairs* 29, no. 9 (Nov.), pp. 26-9
(1978), "The Tektite Problem," *Scientific American* 239, no. 2 (Aug.), pp. 116-25

Olavesen, A.H., *see* Hoyle (1978)

Oparin, A.I., 36
(1953), *The Origin of Life,* tr. Sergius Morgulis (Dover: New York);

orbits, **vii.** 75ff, 11, 14, 16, 22, 24, 27, 32, 38, 40, 47, 63, 74, vii., 147, 167, 173, 178, 198, 215-6, 219, 221, 224-6

Orville, Richard E., 150
(1968), "Photograph of a Close Lightning Flash," *Science* 162 (8 Nov.), pp. 666-7
Ouranos,
also Super-Uranus

Ovendon, Michael W.,

(1972), "Bode's Law and the Missing Planet," *Nature* 239 (27 Oct.), pp. 508-9
et al., (1974), "On the Principle of Least Interaction Action and the Laplacian Satellites of Jupiter and Uranus," *Celestial Mechanics* 8, p . 455-71

Owen, Fraze N., *see* Spangler

Owen, J.R., *see* Canuto

Owen, Tobias, *see* Smith, B.A.

Panagakos, Nicholas & Waller, Peter W., 179, 183
(1974), A New Look at Jupiter: Results from the Pioneer 10 Mission to Jupiter," *NASA News,* no. 74-238, 20 pp. (10 Sep.)
(1979), "Early Findings from Pioneer Venus," *NASA News,* no. 79-13, 22 pp. (5 Feb.)

Pangea(n), 44, 46, 59, 128, 131, 147, 164, 195

P'an Ku, 110, 115, 145

Papoular, R., 96
(1965), "Mobility" in *Electrical Phenomena in Gases,* Eng. ed. (Iliffe: London), Ch. 9, pp. 92-9

Parengo, P.P., *see* Kukarin

Parker, E. N., 32
(1975), "The Sun" in *The Solar System* (Freeman; San Francisco), pp. 27-34

Parkinson, J.H., *see* Clark (1975)

particle, 13, 33-4, 62, 75, 77-9, 96, 123-4, 201-3
is used here as a synonym for electrons, atoms and/ or electron-deficient atoms (ions) which are in motion, such as in an electric discharge, or in a flowing gas or plasma. So viewed, cosmic rays and stellar/ solar wind ions are particles.

Passerini, Pietro, 134
(1978), "Knowledge and Entropy," *Catastrophist Geology* 3, no. 1 (Jun.), pp. 16-28

Patten, Donald Wesley, 128, 189
(1966), *The Biblical Flood and the Ice Epoch* (Pacific Meridian: Seattle)
(1973), *The Long Day of Joshua and Six Other Catastrophes* (Pacific Meridian: Seattle)

Patterson, 183

Peale, S. J., *et al.,* 179
(1979), "Melting of Io by Tidal Dissipation," *Science* 203 (2 Mar.), pp. 892-4

Pearce, G. W., *see* Strangway Pickering, W. H. (1903), *The Moon* (Doubleday: New York); 176 (p. 53)

periastron, 225-6
 means the least separation of the principals in a binary. Similarly, its homologues are perigee and perihelion when orbiting the Earth or the Sun. Elsewhere, the term pericentron is used to describe the closest approach between two bodies in orbit.

Petterson, J.A., *see* Lamers

Phaeton legend, 210-1

phallic, phallus, 166, 173-4

Phillips, J.D., *see* Heirtzler

Philolaos, 73-4

photoshpere, 27-9, 31-3, 93, 116

physical binary system, 23-4
 is here defined to consist of two bodies which are mutually dependent in respect to their orbital revolution about each other. In multiple star systems, which also exist, more than two bodies are in revolution about a common centre-of-motion, often designated as their baricentre.

Pickering, W.H., 178
 (1903), *The Moon,* (Doubleday: New York);

Piltdown Man hoax, 201-2

Pittman, U. J., 98
 (1963), "Magnetism and Plant Growth: I Effect on Germination and Early Growth of Cereal Seeds," *Can. J. Pl. Sci.* 43 (Oct.), pp. 513-8

Plagemann, Stephen, *see* Gribbin
planets, **iv.** 50ff; 11, 13, 22-4, 27, 31, 36, iv., 61-3, 74, 79_81, 97, 105, 108, 114, 117, 122, 131, 146, 155, 167-71, 173, 179, 181-5, 190-1, 194-6, 205-6, 213-8, 221, 224, 226, 230

Plant, A.G., *see* Douglas, J.A.V.

plasma, 28, 33, 68, 93, 214
 is a gas in which the electrons are separated from the electron-deficient atoms. The whole gas contains approximately equal numbers of electrons and ions.

Plass, Gilbert N., 184

(1959), "Carbon Dioxide and Climate," *Scientific American,* offprint no. 82 (repr. from Jul. 1959 issue), 9 pp.

Plato, 66, 74, 158
The Epinomis of Plato, tr. J. Harward (Clarendon: Oxford, 1928)

Plavec, M., 57
(1970), "Rotation in Close Binaries" in *Stellar Rotation,* ed. A. Sletteback (D. Reidel: Dordrecht), pp. 133-46

plenum, v. 61ff; 13, 52-5, 69, 75-78, 80-2, 92, 95-104, 109-113, 115, 117-9, 123, 127-8, 131, 136-7, 143, 146, 154, 157, 166-7, 174, 176, 194, 198, 213, 221
the contents of the sac of Solaria Binaria and later of the Solar System; excluding the distinctly stellar and planetary material in it.

Plinian eruption, 188
is the most violent volcanic eruption known. It is of almost incomprehensible violence such as the eruptions of Stronghyle (believed to have occurred in 1500 BC), of Vesuvius (in AD 79) and of Krakatoa in 1883.

Pliny (Gaius Plinius Secondus), 188
Natural History, Book II. rev., tr. H. H. Rackham (Harvard: Cambridge, 1967)

Plummer, L. N., *see* Sundquist, Ponnamperuma, Cyril & Mack, Ruth (1975), "Nucleotide Synthesis Under Possible Primitive Earth Conditions," *Science* 148 (28 May), pp. 1221-3; 84, fn. 63

polymorphs, 136
are organisms which during their life cycle undergo a transition (metamorphosis) between forms. In some species several forms co-exist within one colony at any moment

polyploids, 105
are species of plants (and sometimes animals) whose chromosome number exceeds twice the basic set of chromosomes (the haploid number) found in the gamete cell (which) produces a new organism by fertilization with an appropriate gamete cell of the opposite gender. It is not uncommon to breed plants with double or four times the original number of chromosomes (euploids).

Ponnamperuma, Cyril & Mack, Ruth, 96
(1975), « Nucleotide Synthesis Under Possible Primitive Earth Conditions, » *Science* 148 (28 May), pp. 1221-3 ;

Poseidon, 161, 167-8
Price, Neville, *see* Norman

primary, 26-7, 45, 48, 52, 57-8, 115-6, 224, 226
    is the major body in a binary system, e. g. the Sun in the Solar System. The companion(s) orbit(s) the primary. In some systems neither object can be called primary.

principals, 23, 31, 52-3, 57-9, 62-3, 67-8, 75, 80-1, 110, 112, 114, 116, 123, 221, 223-4, 226-8
    are the major components in a multiple or binary star system. Referring to Solaria Binaria they would be with time, the Sun and Super Uranus, then after Super Uranus' destruction in a climatic nova eruption, the Sun and Super Saturn. After the Deluge the principals become the Sun and Jupiter whose transactions today dominate motions in the surviving Solar System.

pulsars, 70, 258
    are stars, a significant part of whose observed energy output is not continuous but is emitted as distinct flashes or pulses of electromagnetic radiation. Many pulsars also emit some radiation weakly and constantly, forming a background for the more intensive pulses.

quadrature, 156
    is the angular aspect by which two celestial bodies are observed from a third body to be ninety degrees apart in the sky. An example is the Sun and the quarter-phased Moon as seen from the Earth.

Quaide, William L., *et al.,* 158
    (1970), "Impact Metamorphism of Lunar Surface Materials," *Science* 167 (30 Jan.), pp. 671-2

quantavolution, **xii.** 130ff; **xvii.** 194; 15, 20, 23, 26, 31, 37, 42, 44, 50, 59, 83, 86-7, 90, 94-5, 104-5, 109, 114, 120-1, 194-5, 202-3, 211, 229, 231
    is an abrupt, large-scale change caused by, and affecting one or more spheres such as the astrophere, biosphere, lithosphere, atmosphere, and anthrosphere.

quasar, 71
    is a celestial object which appears "star-like" but is not explainable in terms of the usual stellar properties. Many quasars have a visible "tail" - supposedly a jet of material expelled from the quasar. Often quasars emit anomalous amounts of radio waves.

Quenby, J.J., *see* Coe

radiant genesis, of life, **ix.** 95ff; 46, 131, 133, 137-8

radiation, **vi.** 64ff; 13, 28, 37, 58, 63, 78-9, 96-7, 104, 146, 168, 179, 184, 230-1 is used here to denote electromagnetic waves of any wavelength. It includes, in order of descending wavelength, radiowaves, microwaves, infra-red, visible light, ultra-violet, X-rays, and gamma-rays.

radioactivity of fossils, 13, 135

radiometric, chronology (dating), 19, 130

rain, see deluges

Rampino, Michael R., et al., 94
(1979), "Can Rapid Climatic Change Cause Volcanic Eruptions?," *Science* 206 (16 Nov.), pp. 826-9

Ransom, C. J., 182-3
(1976), *The Age of Velikovsky* (Kronos: Glassboro)
and Hoffee, L. H. (1973), "The Orbits of Venus," *Pensée* 3, no. 1 (Winter), pp. 22-5
also, in *Velikovsky Reconsidered* (Doubleday: Garden City, 1976), pp. 103-9

Rao, C. Baghavender, *see* Matsuoka

Rao, S. Sreedhar, *see* Mitsuoka

Raup, David M., 134-5
(1979), "Size of the Permo-Triassic Bottleneck and Its Revolutionary Implications," *Science* 206 (12 Oct.), pp. 217-8

Rawls, Rebecca, 133
(1980), "Mt St Helens Stirs Chemical Interest," *Chem. & Eng. News* 58 (25 Aug.), pp. 28-9

Reid, G. C., *et al.,* 134
(1976), "Influence of Ancient Solar-Proton Events on the Evolution of Life," *Nature* 259 (22 Jan.), pp. 177-9
Reply by Beland and Russell, *Nature* 263 (16 Sep. 1976), p. 259; also, *Science News* 116 (24 Nov. 1979), p. 356

religion, 109, 116, 131-3, 143, 161-2, 172, 182, 187, 197, 202, 206

resonance, see mutual repulsion

Reynolds, R.T., *see* Peale

Richards, S., *see* Clark (1979)

Ricou, Laurie R., *see* Milton (1978)

rocks, 79, 83, 86, 88, 92, 100, 119, 128, 133, 136, 157-8, 184
Rodabaugh, David J., 105

(1976), "Probability and the Missing Transitional Forms," *Creation Research Quarterly* 13 (Sep.), pp. 116-9

Roosen, Robert G., *et al.,* 94
(1976), "Earth Tides, Volcanos, and Climactic Change," *Nature* 261 (24 Jun.), pp. 680-2

Rose, Lynn, 73-4, 188
(1972), "Could Mars Have Been an Inner Planet?," *Pensée* 2, no. 2 (May), pp. 42-3 also, in *Velikovsky Reconsidered* (Doubleday: Garden City, 1976), pp. 100-2
(1979), "Variations on a Theme of Philolaos," *Kronos* 5, no. 1, pp. 12-46

Ross, John E. & Aller, Lawrence H., 31, 96
(1976), "The Chemical Composition of the Sun," *Science* 191 (26 Mar.), pp. 1223-9

Roussel, René H., 185
"The Reorientation of a Small Temple at Quai es Seboua," *unpubl.*

Routley, Paul McRae, *see* Mihilas

Roy, J.L., *see* Lapointe

Ruark, A. E., *et al.,* 236 (afterglow)
(1927), "Spectra Excited by Active Nitrogen," *J. Opt. Soc. Am.* 14 (Jan.), pp. 17-27

Rudd, P.J., *see* Clark (1979)

Rudeaux, Lucien & de Vaucouleurs, G., 32
(1962), *Larousse Encyclopaedia of Astronomy,* 2nd ed. (Hamlyn: London), 15, fn. 13 (pp. 355 ff.)

Rudnick, Philip, *see* Ruark

Russell, Christopher T., *see* Taylor
et al., (1979), "Pioneer Magnetometer Observations of the Venus Bow Shock," *Nature* 282 (20-27 Dec.), pp. 815-6

Russell, Henry N., *et al.,* 32, 58, 111
(1927), *Astronomy Part II -Astrophysics and Stellar Energy* (Ginn: Boston)

sac, **v.** 58ff; 13, 53-4, 58, 60, 68, 95-7, 100, 102-3, 111, 119, 146-7, 171, 194, 218
in Solaria Binaria, the container of all that can be included in Solaria Binaria, and later on the Solar System; as distinguishable from the medium of space external to it.

Sagan, Carl, *see* Smith, B.A.

salt, 83, 164-6, 170

saltation,
    see Quantavolution

Sanford, Fernando, 92
    (1931), *Terrestrial Electricity* (Stanford University: Stanford), repr. (University Microfilms: Ann Arbor, 1067)

Saturn,
    deities, 72, 114, 132, 143-4, 154, 174-6, 205
    planet, xvi. 158ff; 12, 18, 72, 74, 86, 145, 151, 153-6, 216, 225-6

Saturnian Age, see Age of Saturn

Saul, John M., 121-2
    (1978), "Circular Structures of Large Scale and Great Age on the Earth's Surface," *Nature* 271 (28 Jan.), pp. 345-9
    also, Kellaway, G. A. and Durrance E. M., *Nature* 273 (4 Apr. 1978), p. 75 for a comment extending Saul's sample

Scarf, Frederick L., *see* Gurnett, Taylor

Schaeffer, Claude F.A., 186
    (1948), *Stratigraphie Comparée et Chronologie de l'Asie Occidentale* (Oxford University: London);

Scherer, M., *see* Lemaire

Schmidt, W.M., 162

Schröder, G. A., 67
    (1964), "Breakdown in a Gas Under Extreme Conditions" in *Discharge and Plasma Physics,* ed. S. C. Hayden (The University of New England: Armidale, NSW), pp. 89-106

scientific reception system, 203ff

Second Sun, 161-2, 164

sediments, see rocks

Sekanina, Z., 190
    (1968), "Nongravitational Forces and Comet Nuclei," *Sky and Telescope* 35, no. 5 (May), pp. 282-6

Self, Stephen, *see* Rampino

Seneca, Lucius Annaeus, 167
Natural Questions, tr. Thomas H. Corcoran (Harvard: Cambridge, 1971); 163
(Book III, pp. 27, 29 quoting Berosso's explanations of why the Deluge occurred)

Serson, Paul H., 86
(1980), "Tracking the North Magnetic Pole," *Geos.* (Winter), pp. 15-17

Shamos, Morris H., ed., 219
(1964), *Great Experiments in Physics* (Holt, Rinehart, and Winston: New York)

Shaw, G., *see* Hoyle (Oct. 1977)

Sherrerd, Cris S., 75
(1979), "The Electro-magnetic Circularization of Planetary Orbits," *Kronos* 4, no. 4 (Jun.), pp. 55-8

Shoemaker, Eugene M., see Smith, B. A.
Short, Nicolas M., 48
(1974), "Meteorite Craters," *Ency. Brit. Macro* 12 (Chicago), pp. 48-54; 153. Note one trillion equals 10^{12}

sidereal, 68, 169
measured relative to the stars rather than the Sun.

Sieff, Martin, 187, 189
(1976), "In Defence of the Revised Chronology; An Answer to John Day," *S. I. S. Review* I, no. 1 (Jan.), pp. 11-14
(1977), "Planets in the Bible," *S. I. S. Review* I, no. 4 (Sp.), pp. 17-21, 32
also, *S. I. S. Workshop* 1, no. 4 (Mar. 1978)
(1981), "Assyria and the End of the Late Bronze Age," *S. I. S. Workshop* 4, no. 2 (Sep.), pp. 4-8

Simpson, George Gaylord, 105, 134
(1944), *Tempo and Mode in Evolution* (Columbia University: New York)
(1952), "How Many Species?," *Evolution* 6, p. 342

Simpson, J. A., *et al.,* 184
(1974), "Electrons and Protons Accelerated in Mercury's Magnetic Field," *Science* 185 (29 May), pp. 160-6

Singer, S. Fred, 127, 183
(1970), "How Did Venus Lose its Angular Momentum?," *Science* 170 (11 Dec.), pp. 1196-8
(1967), "Zodiacal Dust and Deep Sea Sediments," *Science* 156 (26 May), pp.1080-3

Sizemore, Warner B., *see* Greenberg

Slavin, J.A., *see* Russell

Smith, Alex G., *see* Lebo
Smith, Bradford A., et al., 179_80
  (1979), "The Galilean Satellites and Jupiter: Voyager 2 Imaging Results," *Science* 206 (23 Nov.), pp. 927-50

Smith, E. v. P. & Jacobs, K. C., 32
  (1973), *Introductory Astronomy and Astrophysics* (Saunders: Philadelphia); 14 (pp. 223 f.)

Soderblom, Laurence, *see* Smith, B.A.

solar, **i.** 22ff; ii. 25ff;
  corona, 27-30, 33, 93, 98
  energy, 96, 118, 123, 134, 212, 230
  flares, 71, 77, 93, 125, 185, 214
  motions, 13, 41-4, 46-8,
  transactions, 61, 84, 102 *et passim*
  wind, 13, 54-6, 83, 93, 125, 184, 195, 212, 214

solar wind, see stellar wind

Somerville, J. M., 67-9
  (1959), *The Electric Arc* (Methuen: London)

sounds, celestial, 19, 81

space-charge sheath, 116, 147
  is a region in which either electrons or electron-deficient atoms predominate and through which electric currents flow. The space-charge limits the current through the sheath. There, electric field strength is not zero.

space infra-charge, 37, 49
  is an electrical property of space itself, not determined by the presence of electrical charges or conductor's residing in that space. The infra-charge is homologous with Paul Dirac's electron theory (1928) which postulated that the vacuum was a sea-of-electrons possessing negative energies. These electrons are not normally detectable but can be prompted into existence (that is, converted into detectable electrons) under certain conditions. The electrons of Dirac's sea affect the energy states of atoms in space. To quote Nobel laureate Leon Cooper (606 fn.): "Thus the vacuum, rather than being an inert void responds to the presence of charges or masses and modifies their behaviour".
  see also Milton (1979), pp 69 and 77, fn.45

Spangler, Steven R., *et al,* 64
  (1977), "Radio Survey of Close Binary Stars," *As. J.* 82, no. 12 (Dec.), pp. 989-97; 47 (p. 989)

species, 79, 99, 104-5, 114, 120, 131, 134-7, 195, 198

specific change ratio,
is a method of comparing the electric charge inherent in a celestial body with some other physical property such as its volume or the number of atoms which it contains. The ratio would thus be expressed in coulombs per cubic metre, coulombs per kilogram, or possibly as excess electrons per kilogram molecular mass (kilomole).

spectrum, 28-9, 38, 43, 45-6, 57, 64, 96, 183, 226

Spencer, Herbert, 162
(1899), *The Principles of Sociology* (Appleton: New York); 158, fn 92 (esp. Ch. 24 and 25)

Sreekantan, B.V., *see* Matsukoa

Shrivastava, B.J., *see* Malin

Stanley, Steven M., 136
(1976), "Stability of Species in Geologic Time," *Science* 192 (16 Apr.), pp. 267-8

stars,
evolution of, 19, 25-7, 33, 37, 39, 40-1, 51, 54-6, 58-60, 63-5, 69, 71, 80, 112-6, 119, 168, 223ff
nearby, 13-4, 19, 23, 42, 44-9

Stecchini, Livio C., 190, 207
(1978), "The Inconstant Heavens" in *The Velikovsky Affair,* rev. (Sphere: London), pp. 80-119; see de Grazia (1967)

Steinhauer, Loren C., *see* Patten (1973)

stellar wind, 27, 30, 113
is the flow of material from a star to the Galaxy. In the electric star the stellar wind exists as one means of the star accumulating charge from the nearly "empty" space which surrounds it. By sending electron-deficient atoms to the Galaxy the star gains electrons relative to the material it contains. From the few stellar winds that have been measured, it seems as if the mass loss increases as the square root of the luminosity. In terms of the electric star model presented here, it is tempting to think that luminosity varies as the square of the star-to-galaxy current. There is some evidence that mass loss is enhanced when a close companion is present (Hutchings).

Stenger, William,
(1978), "Life By Chance?," *The Plain Truth* (Aug.), pp. 13, 15-16

Stewart, John Quincy, *see* Russell, H.N.

Stone, E. C. & Lane, A. L., 179
(1979), "Voyager I Encounter with the Jovian System," *Science* 204 (1 Jun.), pp. 945-8

stones from heaven, see astroblemes

Störmer, Carl,
(1955), *The Polar Aurora* (Clarendon: Oxford)

Strangway, *et al.* 158
(1970), "Magnetic Properties of Lunar Samples," *Science* 167 (30. Jan.), pp. 691-3

Strom, Robert, *see* Smith, B. A.

Strutt, R. J.,
(1911), "A Chemically Active Modification of Nitrogen Produced by the Electric Discharge," *Roy. Soc. (Lon.), Proc.* A 85, pp. 219-29

Suhr, Elmer G., 156
(1969), *The Spinning Aphrodite* (Helios: New York)

Sun, see solar

Sundquist, E. T., *et al.,* 184
(1979), "Carbon Dioxide in the Ocean Surface: The Homogeneous Buffer Factor," *Science* 204 (15 Jun.), pp. 1203-5

Suomi, Verner E., *see* Smith, B.A.

Super Saturn, **xiv.** 158ff; 12, 73, 143-4, 154-6
became the binary partner replacing Super Uranus at the time of the nova explosion which fractured the original companion body into several smaller pieces and Super Saturn.
also,

Super Uranus, **iv.** 46ff; x. 107ff; xiii. 142ff; 11-3, 27, 31, 61-6, 69, 72, 75-6, 80, 82-4, 95, 118-20, 127-30, 159-60, 162, 164, 166, 180, 194, 220, 230

Sutton, Christine, 179
(1979), "Voyage to the Giant Planet," Science 83 (1 Jul.), pp. 213-20

Swaminathan, R., *see* Matsuoka

Synnott, S.P., *see* Morabito

Talbott, David N., 72, 160-1, 166, 174, 183
(1980), *The Saturn Myth* (Doubleday: Garden City)

Talbott, George,
  reply to Forshufvud
  and Ellenberger, Leroy, reply to Morrison, David,

Talbott, Stephen L., *et al.,* (1976), *Velikovsky Reconsidered* (Doubleday: Garden City)

Tanaka, Y., *see* Matsuoka

Tananbaum, H. D. & Hutchings, J. B., 64
  (1975), "Parameters of X-ray Binaries," *NY Acad. Sci., Annals* 262 (15 Oct.), pp. 299-311

Taylor, Jr., Hugh P., *see* Epstein

Taylor, William W. L., *et al.,* 158, 183-4
  (1979), "Evidence for Lightning on Venus," *Nature* 279 (14 Jun.), pp. 614-6

Teichert, Curt (1973), "How Many Species?," *Journal of Paleontology* 30, pp. 967-9

tektites, 126-7

temperatures, 28-9, 32, 38-9, 58, 60, 68, 70-1, 90, 96-7, 146-7, 183-5

tera(amperes), 67, 70
  the prefix tera indicates one million million times the quantity. Tera- is thus a synonym for a multiplier of one billion in Great Britain, and one trillion in the United States. It is, as a measure of current, one million million amperes

thermonuclear fusion, 32, 59
  occurs in a gas of sufficient temperature that its atoms in collision will fuse in significant numbers (see nuclear fusion). A thermonuclear process is purported to provide the power radiated by the stars.

Thor, 172

time, **iii.** 32ff; 19-20, 24, 32, 66, 70, 80, 89, 94-6, 98-9, 105, 111-2, 114, 116, 127-8, 141, 144, 158, 160, 162, 176-7, 179, 182, 185-6, 188-90, 195-7, 202, 207, 216-7, 220, 230
  of lunar eruption, 153

Tinkle, Donald W., 105
  (1974), "Species and Speciation" in *Ency. Brit.,* Macro 17 (Chicago), pp. 449-55

Traill, R.J., *see* Douglas, J.A.V.

Trakhtmann, A.M., *see* Ksanfomaliti
transactive matrix, 34, 213

is a quasi ordered plenum of electrons moving chaotically, which forms a medium through which ions can flow, thereby transmitting an electric currrent. The solar wind electrons form such a matrix, their existence allows the Sun to jettison ions towards the edge of the solar cavity where electrons are readily available.

transmutation, 28, 82, 102
    as used here to transmute means to change the form of, such as from kinetic to potential energy, or to modify the structure of a molecule, crystal, or atom.

tree of life, 72-3, 174

Tresman, Harold & O'Gheoghan, Brendan, 72, 114, 144, 157, 161, 163-4, 166, 171, 174
    (1977), "The Primordial Light," *S. I. S. Review* II, no. 2 (Dec.), pp. 35-40

Treves, A., *see* Maraschi

troposphere, 71, 93
    is the lowest layer of the Earth's atmosphere. It is characterized by the complete mixing of the atoms and molecules of the atmospheric gases by significant vertical winds. The temperature and pressure declines with height in this layer.

Trotter, D.E., *see* Eddy

Turman, B. W., 173
    (1979), "Lightning Detection from Space," *American Scientist* 67 (May-Jun.), pp. 321-9

Tylor, Sir Edward Burnett, 162
    (1903), *Primitive Culture* (John Murray: London), repr. under the title *The Origin of Culture,* Peter Smith (Gloucester MS, 1970); 108, fn. 92 (v. 2, pp. 342 f.)

Ugarte, Patricio, *see* Hesser (1976)

universe, 10, 18-20, 23, 60, 99-101, 109, 143, 172, 195-6, 202-3, 205, 212-3, 215, 218, 230

unseen bodies, 23, 223
    are components in a binary system which remain undetected by direct observation but are implied by some anomalous behaviour of those bodies which are detected.

Uranian Age, see Age of Urania

Uranus (also Uranus minor), *see* Super Uranus

Urey, Harold C. 67, 97, 134
    (1973), "Cometary Collisions and Geologic Periods," *Nature* 242 (2 Mar.), pp. 3, 2-3

Vail, Isaac N., 65
(1905), *Selected Works,* repr. (Annular Publications: Santa Barbara, CA, 1972)

Valentine, James W., 134
(1974), "Temporal Bias in Extinction Among Taxonomic Categories," *Journal of Paleontology* 48 (May), pp. 549-52

van Allen, James, 35
(1975), "Interplanetary Particles and Fields" in *The Solar System* (Freeman: San Francisco), pp. 129-34

van de Kamp, Peter, 23
(1961), "Double Stars," *As. Soc. Pac., Publ.* 73, no. 435 (Dec.), pp. 389-409
(1971), "The Nearby Stars," *An. Rev. As.* Ap. 9, pp. 103-26

van den Heuvel, E.P.J., *see* Lamers, Maraschi

van Flandern, Thomas C., 177, 185
(1978), "A Former Asteroidal Planet as the Origin of Comets," *Icarus* 36, pp. 51-74

Vehrenberg, Hans, 46
Atlas of the Selected Areas (Vehrenberg Publ.: Düsseldorf)

Velikovsky, Immanuel, 87, 115, 121, 125, 158, 161, 176, 181-3, 187, 207, 222, 248
(1950), *Worlds in Collision* (Doubleday: Garden City)
(1952), *Ages in Chaos: I, From the Exodus to King Akhnaton* (Doubleday: Garden City)
(1955), *Earth in Upheaval* (Doubleday: Garden City)
(1969), "Are the Moon's Scars Only 3000 Years Old?," *New York Times,* early ed., 21 Jul.; repr. *Pensée* 2, no. 2 (May 1972)
(1973), "The Pitfalls of Radiocarbon Dating," *Pensée* 3, no. 2 (Sp./ Su,), pp. 12-14, 50
(1978a), "Khima and Kesil," *Kronos* 3, no. 4 (May), pp. 19-23
(1978b), "The Weakness of the Venus Greenhouse Theory," *Kronos* 4, no. 2 (Oct.), pp. 28-32
(1979), "On Saturn and the Flood," Kronos 5, no. 1 (Oct.), pp. 3-11
(1982), *Mankind in Amnesia* (Doubleday: Garden City)

Venus, **xvi.** 181ff; 18, 62, 120-1, 127, 132, 205-6, 221-2, 256, 260, 264, 277, 277, 279, 281, 285,287

Vértes, László, 156
(1965), "'Lunar Calendar' from the Hungarian Upper Paleolithic," *Science* 149 (20 Aug.), pp. 855-6

Vestine, E. H., 86, 88

(1958), "Geomagnetic Field" in *The Earth and its Atmosphere,* ed. D. R. Bates (Basic Books: New York), Ch. 6, pp. 88-96

Veverka, Joseph, *see* Smith, B.A.

visual binary system, 23-4
is a binary system where the component stars are resolvable into separate optical images, that is, the star images are distinguishable.

von Dechend, Hertha, *see* de Santillana

Vsekhsviatskii, S. K., 122
"Indications of the Eruptive Evolution of Planetary Bodies," *unpubl.,* read at the symposium "Velikovsky and the Recent History of the Solar System," McMaster University, Hamilton (18 Jun. 1974), 116

Walborn, Nolan R., *see* Hesser (1976)

Waller, Peter, *see* Panagakos

Walpole, P.H., *see* Simpson, J.A.

Warlow, Peter, 89, 176, 183
(1978), "Geomagnetic Reversals," J. Phys. A10, 11 (Oct.), pp. 2107-30

Warner, Brian & Nather, R. Edward, 62, 113
(1971), "Observations of Rapid Blue Variables -II U Geminorum," *Roy. As. Soc., Mon. Not.* 52, pp. 219-29

Washburn, Sherwood L., 137
"The Evolution of Man," *Scientific American* 239, pp. 194-205

Watson, Alan, 33
(1977), "Whence Cosmic Rays?," *New Scientist* 73 (17 Feb.), pp. 408-10

Webb, Willis, 93
"The Electrical Structure of the Earth," *unpubl.,* read at the symposium "Velikovsky and the Recent History of the Solar System," McMaster University, Hamilton (1974)

Weiner, J. S., 202
(1955), *The Piltdown Forgery* (London: Oxford)

Westfall, Richard S., 217
(1973), "Newton and the Fudge Factor," *Science* 179 (23 Feb.), pp. 751-8

Wang, Y.C., *see* Ness

Whipple, Fred L., *see* Menzel

whistling atmosphere, 81
or whistler, is an electromagnetic wave in the audible frequency range (300 to 30 000 hertz). Its origin is in lightning discharges, and it is propagated along the magnetic field lines (see Hines). Whistlers are today audible only using an amplifier but in the environment of Solaria Binaria they should have been directly audible

White, Fred N., 154
(1974), "Respiration and Respiratory Systems," in *Ency. Brit.* Macro 15 (Chicago), pp. 751-63

Whitten, R., *see* Wolfe

Whyte, Martin A., 99, 134
(1977), "Turning Points in Phanerozoic History," *Nature* 267 (23 Jun.), pp. 679-82

Wickramasinghe, D. T. & Bessell, M. S., 61, 64, 114
(1974), "Gas Streaming in 2UO900-40 and Cyg X-1," *Nature* 251 (6 Sep.), pp. 25-7

Wickramasinghe, N. C., *see* Hoyle

Wickstrom, Conrad E. & Castenholz, Richard W. (1973), 97
"Thermophilic Ostracod: Aquatic Metazoan with the Highest Known Temperature Tolerance," *Science* 181 (14 Sep.), pp. 1063-4

Wigley, T.M.L., *see* Sundquist

Wilkins, H. P. & Moore, Patrick, 158
(1955), *The Moon* (Faber and Faber: London); 154

Windsor, Charles R., *see* Fox

Wiseman, *The Flood Reconsidered;* 156 (p. 46, fn. 9)

Wishaw, Ian Q., *see* Milton (1978)

Wolfe, J., et al., 184
(1979), "Initial Observations of the Pioneer Venus Orbiter Solar Wind Plasma Experiment," *Science* 203 (23 Feb.), pp. 750-2

Wood, John A., 158
(1975), "The Moon" in *The Solar System* (Freeman: San Francisco), pp. 69-77

Wreschner, Ernst, 72
   in conversation with de Grazia, Sept. 1977

Wright, Jr., J.E., *see* Martin

Wright, Kenneth D., 32
   "Observation of Stellar Spectra Related to Extended Atmospheres" in *Extended Atmospheres and Circumstellar Matter in Spectroscopic Binary Systems,* ed. Alen Batten (Int. As. U. Symposium no. 51, May (1973), pp. 117-33

Wrigley, Robert, *see* Quaide

Wyse, Arthur B., 39, 113
   (1934), "A Study of the Spectra of Eclipsing Binaries," *Lick Observatory Bulletin* 17, no. 464, pp. 37-52

X-ray, 61, 64, 71, 78, 113-4, 169, 244

Yallop, B.D., *see* Clark (1979)

Yaniv, A., *see* Heymann

Youngblood, W.W., *see* Blumer

Yukutake, T., 86
   (1967), "The Westward Drift of the Earth's Magnetic Field in Historic Times," *J. Geomag. & Geoel.* 19, no. 2, pp. 103-16

Zasetsey, V.V., *see* Ksanfomaliti

Zeus, see Jupiter

Ziegler, Jerry L., 174
   (1977), *YHWH Star* (Morton, IL)

Zirin, Harold, 30, 53, 71
   (1966), The Solar Atmosphere (Blaisdell: Waltham)

Zwickl, R.D., *see* Krimigis

www.ingramcontent.com/pod-product-compliance
Lightning Source LLC
Chambersburg PA
CBHW020733180526
45163CB00001B/214